SpringerBriefs in Food, Health, and Nutrition

For further volumes:
http://www.springer.com/series/10203

Marco Gobbetti • Raffaella Di Cagno

Bacterial Communication
in Foods

 Springer

Marco Gobbetti
Department of Soil, Plant and Food Science
University of Bari Aldo Moro
Bari, Italy

Raffaella Di Cagno
Department of Soil, Plant and Food Science
University of Bari Aldo Moro
Bari, Italy

ISBN 978-1-4614-5655-1   ISBN 978-1-4614-5656-8 (eBook)
DOI 10.1007/978-1-4614-5656-8
Springer New York Heidelberg Dordrecht London

Library of Congress Control Number: 2012951653

Printed on acid-free paper

Springer is part of Springer Science+Business Media (www.springer.com)

To Camilla and Giuliana so that they always keep the language of their microscopic dreams

Introduction

The levels of bacterial interactions are diverse which makes it rather difficult to distinguish one from the other, especially within complex ecosystems such as foods. Over the last two decades, a number of studies have focused on cell-to-cell communication as a mechanism mainly dedicated to quorum sensing within the microbial population. Findings are accumulating and the potential of cell-to-cell communication is slowly exceeding simple quorum sensing. Some of the most relevant phenotypes, which previously were not fully explained or were thought to be simply orchestrated, are currently believed to be under the control of more sophisticated circuits that inevitably are conditioned by communication within and among bacterial species, and between microorganisms and the human host. Relevant information in this context is also emerging for food-related bacteria.

This book focuses on bacterial communication in foods. First, the main languages used by Gram-negative and -positive bacteria to communicate within and between species are described. The mechanisms of quorum sensing beyond the synthesis of various signaling molecules are highlighted. Once the main bacterial languages are, in part, decoded, the most relevant phenotypes, which are coordinated in a cell density-dependent manner, are reviewed. This mainly includes virulence, biofilm formation, and synthesis of bacteriocins. Next the effects of languages and related phenotypes are translated into food processing and preservation to find the most relevant repercussions. Bacterial communication is described for sourdoughs, yoghurt starter cultures, meat, and vegetables, also giving some insights into quorum quenching mechanisms. As they can be delivered via foods or pharmaceutical preparations, and at least in some aspects have an impact on human health, one chapter focuses on bacterial quorum sensing mechanisms that occur at the gastrointestinal level between probiotic bacteria and other intestinal inhabitants, and between probiotics and the host. The new perspective, which emerges from the paradigm shift in how the bacterial population has to be perceived and controlled, is given in the concluding chapter of this book.

Raffaella Di Cagno is a researcher at the Department of Soil, Plant and Food Science at the University of Bari Aldo Moro, Italy, specializing in Food Microbiology. She is the author of 75 articles published in international journals, which deal with food microbiology.

Marco Gobbetti is full professor of Food Microbiology in the Department of Soil, Plant and Food Science at the University of Bari Aldo Moro, Italy. He is the author of approximately 320 articles, the majority of which are published in international journals. In recent years, bacterial communication has been one of his main research topics.Bacterial Communication in Foods

Contents

Chapter 1
The Language

1.1 Introduction

For a long time, microorganisms were believed to exist as individual cells, whose primary aims were finding nutrients and multiplication. Today, it is clear that microorganisms perform coordinated activities, which previously were restricted to multicellular organisms. Microbial communities exhibit all the hallmarks of a complex and social life. The term socio-microbiology was coined, which was aimed at exploiting collective microbial behavior [1]. The levels of microbial interaction are diverse, mainly positive (e.g., proto-cooperation, symbiosis, commensalism) or negative (e.g., competition, amensalism, parasitism). The highest and most sophisticated form of interaction or social behavior is coordinated microbial communication, which, in most cases, is cell density dependent. Apart from direct cell-to-cell contact, the synthesis of small diffusible chemicals, probably, offers the most efficient strategy for communication among microorganisms.

Under this social context, it is not easy to distinguish between cell-to-cell communication and the many other examples of cell interaction. Although in its infancy, examples of communication are already reported for bacteria, yeasts, and moulds. To date, cell-to-cell communication is mainly described among bacteria (ca. 3,300 hits are found using the main search databases). The mechanisms that regulate bacterial communication are, for the most part, common to various habitats, but foods and beverages are ecosystems where the ecology of communication shows intrinsic features that have indubitable repercussions on the quality of products and human health. The study of cell-to-cell communication in food-related bacteria is, therefore, becoming an extremely attractive area of food microbiology, which is generating a significant paradigm shift in terms of how the microbial population is perceived and controlled.

M. Gobbetti and R. Di Cagno, *Bacterial Communication in Foods*,
SpringerBriefs in Food, Health, and Nutrition, DOI 10.1007/978-1-4614-5656-8_1,
© Marco Gobbetti and Raffaella Di Cagno 2013

1.2 Signals from Bacteria

Foods and beverages are complex ecosystems where the chemical languages (signal, cue, or communication) between interacting bacteria may diffuse throughout the microbial community. To be classified as true communication, a signal compound has to be created for transmitting information, has to be perceived by others, and has to engender a response in the receiver [2]. In the literature, words such as language and behavior are frequently used to depict the mechanism mainly dedicated to sensing the quorum of the microbial population (quorum sensing). Simplifying this concept, language and cross-talk between bacteria, between bacteria and other eukaryotic microorganisms, and between bacteria and animal, human, or plant hosts should determine their behavior (e.g., beneficial or pathogenic phenotypes).

One major concern was and is to understand and to decode this language. Given the large number of extracellular metabolites, the chemical diversity of known quorum sensing signals is likely to represent the tip of the iceberg. The main quorum sensing processes are summarized in Fig. 1.1. Among the many different signaling languages, the well-known words used by Gram-negative bacteria are the N-acyl-L-homoserine lactones (AHL). Languages based on the synthesis of

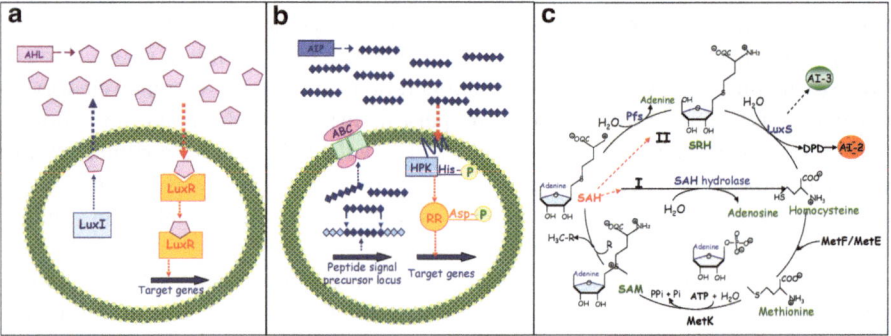

Fig. 1.1 Schematic representation of the main bacterial quorum sensing signal transduction circuits. (**a**) Canonical LuxI/LuxR quorum sensing system of Gram-negative bacteria. The LuxI-like protein is the autoinducer synthase, which is responsible for the formation of the specific N-acyl-L-homoserine lactone (*AHL*) autoinducer molecule (*pink pentagons*). The autoinducer freely diffuses through the bacterial cell envelope and accumulates at high cell density. When a sufficiently high concentration of the signal is achieved, the cytoplasmic receptor protein LuxR binds its cognate autoinducer. The LuxR-autoinducer complex also binds to target gene promoters and activates transcription (Adapted from [71]). (**b**) Oligopeptide-mediated quorum sensing of Gram-positive bacteria. A specific precursor peptide (*large sky blue-violet bars*) is synthesized, modified, and processed. An ABC exporter complex secretes the mature oligopeptide signal (*short violet bars*). The oligopeptide accumulates as the bacterial cell density increases. A two-component signal transduction system is responsible for detection of the signal and relaying the information into the cell. Signal transduction occurs via a phosphorylation cascade (*P*). The model shows a sensor histidine protein kinase (*HPK*) and response regulator (*RR*) containing histidine and aspartate residues, which correspond to the sites of phosphorylation. The signal transduction cascade results in an alteration of the gene expression of specific target outputs (Adapted from [63]). (**c**) Synthesis of the autoinducer-2 (*AI-2*) via LuxS. During AI-2 biosynthesis, the transfer of the methyl group from

post-translationally modified peptides are confined to Gram-positive bacteria. The universal chemical lexicon, shared by both Gram-negative and -positive bacteria, involves the synthesis of the autoinducer-2 (AI-2), through the activity of the LuxS (S-ribosyl homocysteine lyase) enzyme. A new bacterial signal, dependent or not on LuxS and termed autoinducer-3 (AI-3), was decoded in some species of Gram-negative bacteria, which are resident in the human gastrointestinal tract. The language based on AI-3 seems to be involved in cross-talk between bacteria and the human host (e.g., interkingdom communication, see Sect. 4.3).

1.3 N-acyl-L-Homoserine Lactones

Various Gram-negative bacteria produce cell-permeable AHL at a low basal rate. If these signals are allowed to accumulate, they bind cognate transcriptional regulators and act as autoinducing signals. Since the concentration of autoinducers often mirrors the local population density, they act as a sort of census to regulate gene expression in a population-dependent manner. The general paradigm is that species-specific quorum sensing in Gram-negative bacteria is mediated by AHL, as their primary autoinducers [3]. Most of the AHL-producing bacteria synthesize multiple AHL. More than 20 different AHL are known. They share a common homoserine lactone (HSL) ring, which is un-substituted at the β- and γ-positions, and N-acylated at the α-position. The length of the acyl chain varies from C4 to C18. A ketone is frequently found at C3, which is sometimes reduced to a hydroxyl group. The acyl chain is branched or, in some cases, unsaturated. The cell membrane is permeable to AHL, and thus these molecules are secreted outside the cell and accumulate in the surrounding environment. At low cell densities, AHL passively diffuses out of cells down a concentration gradient, while at high cell densities, AHL accumulates at an intracellular concentration (ca. 10 nM) equivalent to the extracellular level [4].

First, the mechanism of quorum sensing was described in *Vibrio fischeri*, a bioluminescent bacterium living as a symbiont in specialized light organs of the Hawaiian bobtail squid *Euprymna scolopes* and fish *Monocentris japonicus* [5]. The acyl-HSL, 3-oxo-hexanoyl-HSL (3-oxo-C6-HSL), synthesized by *V. fischeri* is shown in Fig. 1.2. Symbiosis is coordinated through a three-component quorum sensing circuit, which is made up of an autoinducer signal, and its synthase (LuxI) and receptor (LuxR). S-adenosylmethionine (SAM) and an acylated acyl carrier protein

Fig. 1.1 (continued) S-adenosylmethionine (*SAM*) to its substrate leads to the synthesis of S-adenosyl homocysteine (*SAH*). SAH is toxic for the cell and it is removed following two routes: (*I*) one-step conversion through SAH-hydrolase activity or (*II*) two-step conversion via the activity of Pfs and LuxS enzymes. The Pfs nucleosidase enzyme hydrolyzes adenine from SAH to form S-ribosylhomocysteine (*SRH*). LuxS acts on SRH to synthesize 4,5-dihydroxy-2,3-pentanedione (*DPD*) and homocysteine. DPD undergoes further rearrangements to synthesize the active AI-2 molecule. *Dotted arrow* indicates the hypothetic release of the autoinducer-3 (*AI-3*). MetF, methylene tetrahydrofolate reductase; MetE, cobalamin-independent methionine synthase; MetK, methionine adenosyltransferase (Adapted from [69])

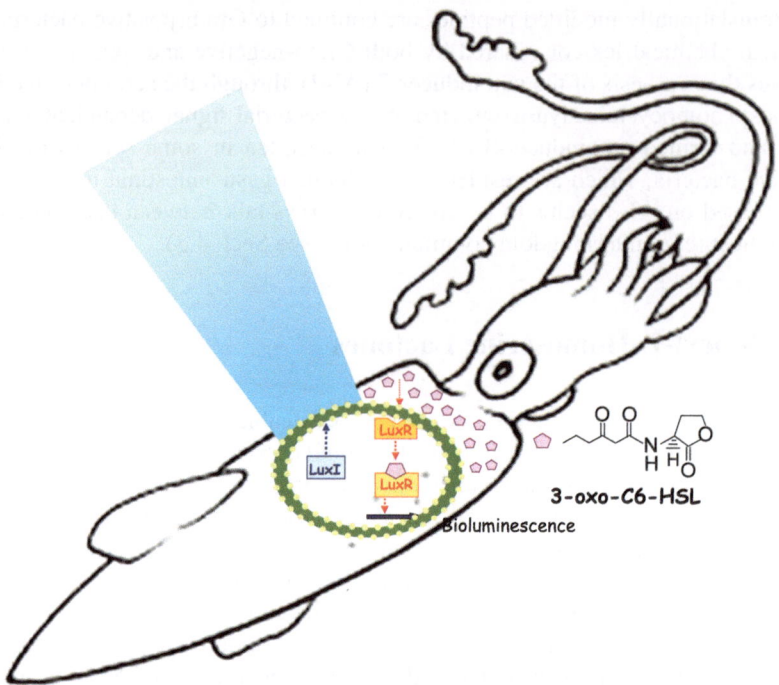

Fig. 1.2 Schematic representation of a three-component quorum sensing circuit and its role in the symbiotic relationship between *Vibrio fischeri* and the Hawaiian bobtail squid (*Euprymna scolopes*). *V. fischeri* inhabits the light organ of the squid, and uses quorum sensing to bioluminesce at high cell densities. In turn, the squid uses this bioluminescence for camouflage and other processes. The three-component quorum sensing circuit includes an autoinducer signal 3-oxo-hexanoyl-HSL (*3-oxo-C6-HSL*), its synthase (*LuxI*) and receptor (*LuxR*)

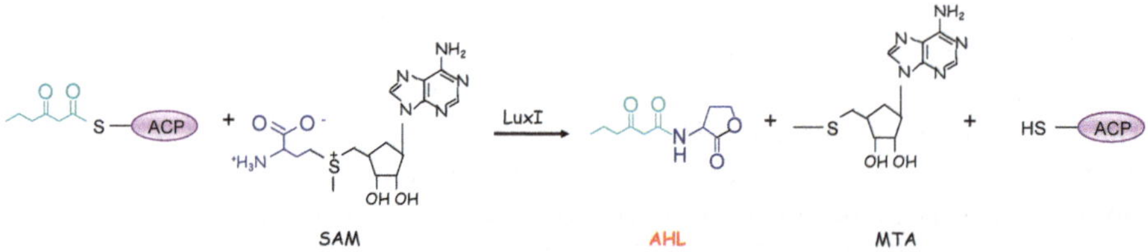

Fig. 1.3 Chemical reaction catalyzed by the autoinducer synthase LuxI to synthesize N-acyl-L-homoserine lactones (*AHL*). All known LuxI-type AHL synthases utilize S-adenosyl methionine (*SAM*) as a substrate; product diversity comes from the different acylated acyl carrier proteins (*ACP*) that are used as the second substrate. MTA, methylthioadenosine (Adapted from [6])

(ACP) are the substrates of LuxI [6]. In general, two distinct chemical reactions are required to form an AHL: (1) the acyl transfer from ACP to the amino group of SAM; and (2) the lactonization of SAM, with the concomitant excretion of S-methylthioadenosine (MTA) (Fig. 1.3). Usually SAM acts as a methyl group donor, therefore, its role as a source of amino acid during synthesis of AHL is physiologically unusual. Once 3-oxo-C6-HSL is synthesized by *V. fischeri*, it readily diffuses into the organism's environment. The concentration of the signal increases with the size of the bacterial population. Once the intracellular threshold of 3-oxo-C6-HSL is reached, it binds to its cognate cytoplasmic receptor protein, the LuxR transcription factor. The transcriptional activator LuxR and 3-oxo-C6-HSL direct the population density responsive transcriptional activation of the *luxICDABEG* operon, which is involved in the symbiosis and bioluminescence processes [7]. Luminescence also depends on 3':5'-cyclic adenosine monophosphate (AMP) and the GroESL chaperone, which influence the synthesis and activity of LuxR, respectively. The elucidation of this canonical quorum sensing system in *V. fischeri* demonstrated that a single chemical signal could initiate a set of complicated binding events that controlled important bacterial functions.

All AHL-type quorum sensing circuits, which were subsequently characterized in other bacteria, contain homologues to LuxI and LuxR regulatory proteins. The number of AHL that are defined as quorum sensing molecules largely exceeds the number of the corresponding synthases. Sequence homology among AHL synthases suggests that they are structurally similar and, probably, follow similar chemical mechanisms [6]. The structure of three AHL synthases was elucidated through X-ray crystallography: EsaI from *Pantoea stewartii*, which catalyzes the formation of 3-oxo-hexanoyl homoserine lactone [8], LasI from *Pseudomonas aeruginosa*, whose product is 3-oxo-dodecanoyl homoserine lactone [9], and TofI from *Burkholderia glumae*, which catalyzes the formation of octanoyl homoserine lactone (C8-HSL) [10]. AHL synthases typically exhibit strict, but not absolute, substrate specificity. For example, the best substrate for RhlI, another AHL synthase from *P. aeruginosa*, is butyryl-ACP, but it may also catalyze the slow formation of N-hexanoyl homoserine lactone [11]. The structures of EsaI and TofI suggest that substrate specificity is determined by the size of the tunnel, where the acyl chain of acyl-ACP binds. Substitution of two residues in the AHL synthase of *Erwinia carotovora* altered its specificity: N-(3-oxooctanoyl)-L-homoserine lactone, rather than the normal product N-(3-oxohexanoyl)-L-homoserine lactone, was synthesized [12]. Recently, it was discovered that some bacteria produce *p*-coumaroyl homoserine lactone when growing in the presence of *p*-coumarate [13]. Since *p*-coumarate is not a bacterial metabolite but a component of lignin, the main suggestion was that bacteria rely on the plant host to obtain the side chain needed to synthesize the signaling molecule. Although not already demonstrated, bacteria may express the enzymes to convert *p*-coumarate into *p*-coumaroyl-ACP.

Table 1.1 lists some food-related bacterial species, which are known to possess LuxI/LuxR like quorum sensing systems. As mainly shown using mutant strains, generally, other metabolic functions than quorum sensing are regulated through these systems. Five non-Lux quorum sensing regulated proteins were identified in mutants of *V. fischeri*, which were defective for LuxR and AHL signals [14]. Their synthesis mainly occurs at high density population, requires both LuxR and 3-oxo-C6-HSL, and is inhibited by C8-HSL at low density population. Genes encoding two of the five quorum sensing regulated proteins were characterized: *qsrP* directs the synthesis of a novel periplasmic protein and *ribB* expresses the 3,4-dihydroxy-2-butanone 4-phosphate synthase, a key enzyme to synthesize riboflavin, which is the

Table 1.1 LuxI/LuxR like quorum sensing systems of some food-related Gram-negative bacteria: regulatory proteins, autoinducer identity, and target genes and functions

Bacterium	Regulatory proteins	Autoinducer identity	Target genes and functions
Vibrio fischeri	LuxI/LuxR	N-(3-Oxohexanoyl)-HSL	*luxICDABE* Bioluminescence
Burkholdeira cepacia	CepI/CepR	N-Hexanoyl-HSL and N-octanoyl-HSL	*cep* box Biofilm formation, swarming motility, and production of extracellular protease
Enterobacter agglomerans	EagI/EagR	N-(3-Oxohexanoyl)-HSL	Unknown
Erwinia carotovora	(1) ExpI/ExpR	N-(3-Oxohexanoyl)-HSL	(1) *rsm* Exo-enzyme synthesis
E. carotovora	(2) CarR/CarI	N-(3-Oxohexanoyl)-HSL	(2) *carBCDEFG* Carbapenem antibiotic synthesis
Escherichia coli	?/SdiA	?	*ftsQAZ* Cell division and chromosome replication
Pseudomonas aerofaciens	PhrI/PhrR	N-Hexanoyl-HSL	*phz* Phenazine antibiotic biosynthesis
Pseudomonas aeruginosa	(1) LasI/LasR	(1) N-(3-Oxododecanoyl)-HSL	(1) *lasA, lasB, aprA, toxA* Exo-protease virulence factors and biofilm formation
P. aeruginosa	(2) RhlI/RhlR	(2) N-Butyryl-HSL	(2) *rhl*AB Rhamnolipid synthesis; *las*B, and *apr*A, stationary-phase sigma factor RpoS synthesis and production of the secondary metabolites
Salmonella Enteritidis	?/SdiA	?	*rck* Resistance to complement killing, *srgA, srgB,* and *srgC* virulence plasmid synthesis
Serratia liquefaciens	SwrI/SwrR	N-Butanoyl-HSL	*bsmA* and *bsmB*, biofilm formation *lipB*, secretion of extracellular lipase, metalloprotease, and S-layer protein
Yersinia enterocolitica	YENI/YENR	N-Hexanoyl-HSL, N-(3-oxohexanoyl)-HSL	Unknown

HSL homoserine lactone

precursor of the luciferase substrate. Compared to wild type, Δ *qsrP* and Δ *ribB* mutants did not exhibit distinct phenotypes but were less successful in colonizing *E. scolopes*.

The acyl-HSL quorum sensing of the opportunistic pathogen and occasionally food-related *P. aeruginosa* represents one of the best understood systems of cell-to-cell communication. Quorum sensing is controlled by a powerful signaling triumvirate, which directs the bacterial group behavior [15] (Fig. 1.4). The system is highly intertwined with other cellular pathways, rendering it responsive to a multitude of environmental signals. Despite this complexity, many of its features are solely explained through the compact circuitry of two synthases and three receptors. LasR, RhlR, and QscR are the receptors homologues to LuxR. A fourth receptor, PqsR, is a LysR-like receptor that recognizes the *Pseudomonas* quinolone signal PQS, and it is intimately connected with the other three [16, 17]. LasR, RhlR, and QscR are collaborative components of the quorum sensing system, which is designed for adaptability and persistence in unforgiving environments. LasR positively regulates the RhlR system, and together these two systems are responsible for regulating PqsR. In turn, QscR represses LasR and RhlR [17, 18]. LasI and RhlI synthase generate the autoinducers 3-oxo-dodecanoyl homoserine lactone (OdDHL) and butanoyl homoserine lactone (BHL), which are recognized by LasR and RhlR receptors, respectively. The OdDHL:LasR and BHL:RhlR complex induce a variety of genes of *P. aeruginosa*, including those for the synthesis of the rhamnolipid biosurfactant [17]. These two systems (LasI/R and RhlI/R) control ca. 6 % of the *P. aeruginosa* genome [19]. The hierarchal relationship between LasR and RhlR is largely dependent on empirical conditions, which may be manipulated to favor virulent phenotypes, independently from the functional quorum sensing system [17]. The third LuxR-type receptor of *P. aeruginosa*, QscR, diverges from LasR and RhlR. QscR is an orphan receptor that lacks the associated LuxI-type enzyme [17]. Without a cognate synthase, the activity of QscR relies on OdDHL, the signal synthesized by LasI. A *P. aeruginosa* mutant, which was lacking in *qscR*, showed hyper-virulence in vivo, which suggested that QscR acts as a repressor of quorum sensing [20]. The discovery of this mechanism originated the term QscR, as the quorum sensing control repressor protein. Contrarily to LasR and RhlR, QscR binds numerous AHL with a range of acyl tail lengths and oxidation states at the 3-position. The speculation was that QscR plays a role not only in the intra- but also in the interspecies communication [21]. As an orphan receptor with relatively high ligand promiscuity, QscR is considered an attractive candidate to control the virulence of *P. aeruginosa*. Additional components, which are essential to the quorum sensing system of this bacterium, are constantly identified. A recent example is the resistance-nodulation-division-type efflux pump MexAB-OprM, which selects the access to the specific OdDHL language rather than to other acyl-HSL [22]. Nevertheless, the relationships between the already established factors and the additional components of the quorum sensing system are not fully understood.

Erwinia carotovora is a phytopathogen that causes soft rot in a variety of plant products (e.g., potato, carrot, and celery) through plant cell wall-degrading enzymes such as pectate lyases, polygalacturonase, cellulase, and protease. The synthesis of

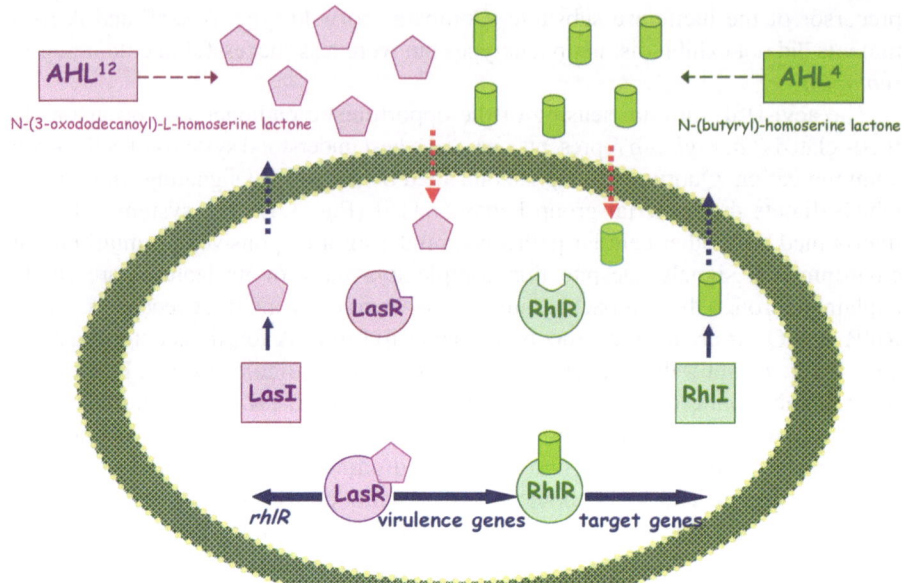

Fig. 1.4 Hierarchical quorum sensing in *Pseudomonas aeruginosa*. LasI, autoinducer synthase homologous to LuxI; AHL[12], N-(3-oxododecanoyl)-L-homoserine lactone; LasR, transcription factor homologous to LuxR; RhlI, autoinducer synthase homologous to LuxI; AHL[4], N-(butyryl)-homoserine lactone; RhlR, transcription factor homologous to LuxR; and *rhlR*, gene encoding RhlR. For the quorum sensing mechanism see the text (Adapted from [63])

these enzymes by only a few cells would not have an effect on plant tissue, and it would activate the plant phytodefense mechanisms. Therefore, *E. carotovora* uses quorum sensing, which ensures that the synthesis of enzymes does not occur until sufficient bacterial numbers are achieved [23]. This regulation relies on ExpR/ExpI, which are homologues to LuxR/LuxI. Although pectinases are the primary determinants of virulence, several ancillary factors that augment the bacterial virulence were also identified. Bacterial motility is one of these factors [24]. Flagellum formation and bacterial movement are regulated in many enterobacteria, including *E. carotovora*, via FlhDC, the master regulator of flagellar genes, and FliA, the flagellum specific σ factor. As shown using *fliC* and *motA* mutants, the motility of *E. carotovora* is positively regulated by AHL and negatively regulated by RsmA, a post-transcriptional regulator [24]. Nonmotile mutants showed a decreased capacity to cause soft-rotting disease in Chinese cabbage. Another mechanism of quorum sensing (CarR/CarI) regulates the synthesis of the antibiotic carbapenem by *E. carotovora* [25]. When sufficient AHL is synthesized, the CarR receptor is activated, which induces the expression of the carbapenem genes. Due to the very complex microbiota of plants, it seems that *E. carotovora* counteracts microbial competition by coordinating the synthesis of carbapenem together with that of tissue-macerating enzymes.

Yersinia enterocolitica, an intestinal bacterium also conveyed by foods (e.g., fresh meat, fish, and vegetables), is the most common *Yersinia* species among pathogens in humans [26]. The YENR/YENI locus (homologues to LuxR/LuxI) and the signals N-hexanoyl-L-homoserine lactone (C6-HSL) and N-(3-oxohexanoyl)-L-homoserine lactone (3-oxo-C6-HSL) constitute the quorum sensing system of *Y. enterocolitica* [27]. Compared to other Gram-negative bacteria, *Escherichia coli* and *Salmonella enterica* subsp. *enterica* serovar *enteritidis* (Salmonella Enteritidis) seem to have an incomplete set of LuxI/LuxR homologues [28]. SdiA is the sole LuxR-type receptor found in *E. coli* and Salmonella Enteritidis. No *luxI* homologues, or any other type of AHL synthase, were found in the complete genome sequence. SdiA is also considered as an example of bacterial receptor that detects signals of other microbial species. Exceptions to the LuxI/LuxR and AHL paradigm are not unusual in Gram-negative bacteria. Indeed, genomic data indicate the presence of putative signaling molecules such as peptides, cyclic dipeptides and esters, and several transporters, which specifically allow the diffusion of the signals [29]. A system, which uses a protease to release the signaling peptide, was discovered in the genus *Providencia* [28]. The 3-hydroxypalmitic acid methyl-ester seemed to regulate the virulence of *Ralstonia solanacearum*, a soil-borne phytopathogen that causes wilting diseases of many important crops [30]. Diketopiperazines were discovered in *Pseudomonas fluorescens*, *Pseudomonas alcaligenes*, and *Enterobacter agglomerans* [31].

Since the discovery in *V. fischeri* over 30 years ago, our understanding of the quorum sensing mechanisms in Gram-negative bacteria has rapidly expanded. The complexity of the circuitries, which control the collective cell behavior, is also emerging.

1.4 Autoinducing Peptides

The signaling molecules, the mechanism of their synthesis, and the secretion and detection apparatus used by Gram-positive bacteria differ from those of Gram-negative bacteria (Table 1.2). Gram-positive bacteria use a ribosomally generated oligopeptide called the autoinducing peptide (AIP, or peptide pheromone) as the communication signal. The gene for AIP often flanks a two-component regulatory system (2CRS) gene cassette [32]. The system that includes AIP in addition to 2CRS is termed the three-component regulatory system (3CRS). The 2CRS, consisting of a membrane-bound sensor histidine protein kinase (HPK) and a cognate cytoplasmic response regulator (RR), is the major tool for signal transduction across cell membranes in bacteria [33]. HPK and RR contain characteristic domains, termed transmitters and receivers, respectively. In the simplest circuit, HPK contains a C-terminal transmitter module, which is preceded by an N-terminal signal input domain with an invariant auto-phosphorylated histidine. RR contains an N-terminal receiver module, which is followed by a C-terminal signal output domain with an invariant aspartate residue located at the center. In the 3CRS process, AIP, which is secreted by the dedicated

Table 1.2 Quorum sensing systems of some food-related Gram-positive bacteria: signaling molecules and functions

Bacterium	Signaling molecules	Functions
Bacillus subtilis	ComX, CSF subtilin	Competence/sporulation and lantibiotic synthesis
Carnobacterium maltaromaticum	AMP-like peptide pheromone (CS)	Class II bacteriocin synthesis
Carnobacterium piscicola	AMP-like peptide pheromones (CbnS, CbaX)	Class II bacteriocin synthesis
Enterococcus faecalis	GBAP, and "CyILs", AMP-like peptide pheromone (EntF)	Virulence and Class II bacteriocin synthesis
Lactobacillus plantarum	LamD558 AMP-like peptide pheromone (PlnA)	Exopolysaccharides synthesis, cell membrane proteins and Class II bacteriocin synthesis
Lactobacillus sakei	AMP-like peptide pheromone (SppIP)	Class II bacteriocin synthesis
Lactococcus lactis	Nisin	Lantibiotic synthesis
Staphylococcus aureus	AIP, AgrD (agr system)	Virulence
Streptococcus pneumoniae	CSP, BlpC	Competence and virulence

CSF competence and sporulation factor, *AMP* antimicrobial peptide, *CbnS and CbaX* carnobacteriocins, *GBAP* gelatinase biosynthesis-activating pheromone, *"CyILs"* cytolysin, *EntF* enterococin, *PlnA* plantaricin A, *SppIP* sakacin, *AIP* autoinducing peptide, *CSP* competence-stimulating peptide

ATP-binding-cassette, accumulates outside the cell. When its concentration reaches the quorum, AIP triggers HPK by binding to its N-terminal sensor domain. The HPK auto-phosphorylates its own histidine residue and transfers the phosphate group to the aspartate of RR. Finally, the phosphorylated RR activates gene transcription. Contrarily to AHL molecules, which diffuse freely across the outer and inner membranes and require an internalization to trigger the corresponding response, the peptide pheromone model does not include this step. The sensor protein is located on the outer surface of the cytoplasmic membrane.

A variety of phenotypes are controlled by 3CRS: (1) synthesis of bacteriocins in *Carnobacterium piscicola* [34], *Lactobacillus sakei* [35], *Lactobacillus plantarum*, and *Enterococcus feacium* [36]; (2) conjugal transfer of plasmids in *Enterococcus faecalis* [37]; (3) genetic competence in *Streptococcus pneumoniae* [38] and *Bacillus subtilis* [39]; (4) sporulation in *B. subtilis* [40]; (5) expression of virulence factors in staphylococci [41]; (6) biofilm formation; and (7) stress responses [42] (see Chap. 2). Although quorum sensing systems may control the behavior of numerous Gram-positive bacteria via 3CRS, the complexity of the signals that they perceive still remains to be fully defined.

Most of the accessory genes (virulon), which are involved in pathogenesis by *Staphylococcus aureus*, encode proteins that are either displayed on the bacterial surface or released into the surroundings. These proteins enable the organism to evade the host defenses, to adhere to and to degrade cells and tissues, and to spread within the host. The global regulation of the staphylococcal virulon is via the gene regulator *agr* locus, a quorum sensing system that controls the expression of most of the exo-protein genes [43]. As shown in animal models, the attenuation of the virulence occurred in *agr* mutants [43]. Other pleiotropic mutants, showing the mutation of virulence gene regulators such as *sarA* [44], *sarS* [45], and *rot* [46], supported the concept of global regulation of the virulence. These regulatory systems sense and integrate various extra- and intracellular inputs such as cell density, energy availability, and environmental signals, which determine the synthesis of exo-proteins only when they are required. Information on environmental conditions is read via signal receptors, which for *S. aureus* is the primary regulatory circuit to express the virulon. In addition to 2CRS *agr*, three other distinct CRS are involved: *sae*, *srh*, and *arl*. All four gene regulators represent one-quarter of the putative CRS identified in the genome of *S. aureus*. The *agr* locus of *S. aureus* consists of two divergently transcribed operons, RNAII and RNAIII [47] (Fig. 1.5). The RNAII operon contains the *agrBDCA* genes that encode the signal transducer (AgrC, homologous to HPK) and the response regulator (AgrA, homologous to RR). During growth, a small (7–9 amino acids in length) extracellular AIP, derived from a dedicated pro-peptide (AgrD) and encoded by *agr*D, is secreted and accumulates. Upon reaching the threshold concentration (ca. 10 nM), AIP binds to and triggers activation of the AgrC signal transducer, which auto-phosphorylates and, in turn, leads to the phosphorylation of the AgrA response regulator [48]. Phosphorylated AgrA stimulates the transcription of RNAIII and RNAII, and up-regulates the expression of numerous exo-proteins as well as that of the *agrBDCA* locus [47]. The latter leads to the rapid increase of the synthesis and export of AIP. At the second regulatory locus, the *sar* gene product (SarA) functions as a regulatory DNA-binding protein for inducing the expression of both RNAII and RNAIII operons. The *agr* quorum sensing system of *S. aureus* also represses the synthesis of several surface adhesins (e.g., fibrinogem- and fibrinectin-binding proteins) that mediate the contact with the host matrix [49].

Analogous to the *agr* system of *S. aureus*, a similar 3CRS (regulator *fsr*) exists in *E. faecalis* [50]. This locus includes homologous to HPK, FsrC, to RR, FsrA, and a putative AgrB-like processing enzyme, FsrB. All genes of the *fsr* operon are important for the synthesis of virulence factors. Contrarily to the *agr* system, the AIP of *E. faecalis* is derived from the C-terminus of FsrB. Another quorum sensing circuit was found in clinical isolates of *E. faecalis*, which produces the exo-toxin called cytolysin. Two regulatory proteins cylR1 and cylR2, which lack homologues of known function, work together to repress the transcription of the above cytolysin genes. De-repression occurs at specific cell density, when one of the cytolysin subunits reaches the extracellular threshold concentration [51].

Lactic acid bacteria comprise a diverse group of Gram-positive bacteria, routinely used for the manufacture of fermented foods and beverages, but also largely found as natural inhabitants of the gastrointestinal tract of humans and animals. Lactic acid bacteria use various sensory strategies to allow an efficient colonization and adaptation to changing environmental conditions. The locus of the *L. plantarum*

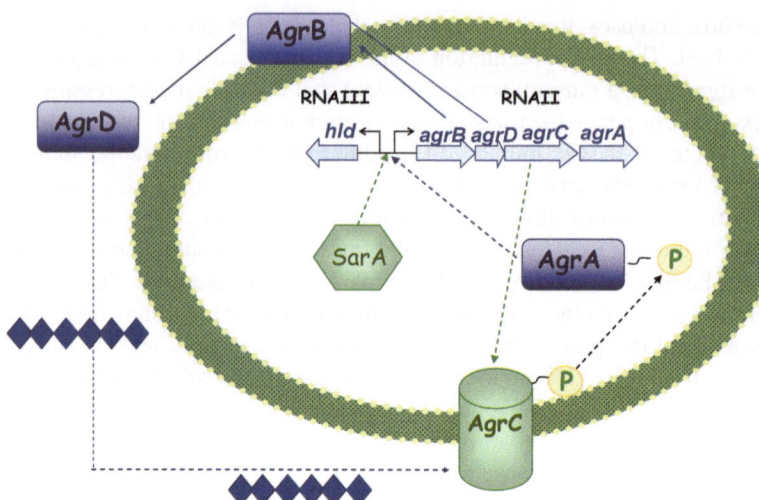

Fig. 1.5 Quorum sensing in *Staphylococcus aureus*. One operon RNAII (*agrBDCA*) encodes proteins responsible for generating and sensing the peptide signal molecule (*AgrD*), and the other operon encodes δ-hemolysin and RNAIII. AgrB, membrane-associated protein; AgrC, transmembrane associated signal transducer; AgrA, response regulator; and SarA, DNA-binding protein. For the quorum sensing mechanism see the text (Adapted from [15])

WCFS1 genome, which showed homology to the staphylococcal 3CRS *agr*, was analyzed and designated as *lam* [52]. The analysis of the response regulator-defective mutant (*ΔlamA*) showed that *lam* regulates the adherence of *L. plantarum* to the glass surface. The microarray analysis confirmed that *lamBDCA* is auto-regulatory and showed that *lamA* is responsible for the synthesis of surface polysaccharides, and cell membrane and sugar fermentation proteins. A cyclic thiolactone pentapeptide, which possesses a ring structure similar to that of the staphylococcal AIP, was identified and designated as LamD558.

Convergent pathways regulate the quorum response of *B. subtilis*. This is mediated via a secreted 10-amino-acid-modified peptide, ComX pheromone, which activates ComP (homologous to HPK) that, in turn, stimulates ComA (homologous to RR). A competence and sporulation factor (CSF), quorum sensing pentapeptide, which derives from the precursor phrC gene product, also stimulates ComA activity (see Sect. 4.3). CSF is imported into the cells by an oligopeptide permease and inhibits the putative aspartyl-phosphate phosphatase RapC, which negatively regulates ComA. The expression of over 20 genes seems to be under the direct control of this signaling pathway, and the expression of over 150 additional genes, including those responsible for competence, are controlled indirectly. Overall, these controlled genes regulate and enhance survival, growth, and colonization under conditions of crowding [53].

Exceptions to 3CRS were also shown. For instance, small signaling molecules, known as γ-butyrolactones, which are structurally similar to Gram-negative AHL and function in a cell density-dependent manner to elicit the synthesis of antibiotics, were identified within the genus *Streptomyces* [54].

1.5 *LuxS/autoinducer-2*

The quorum sensing circuit of bioluminescent *V. harveyi* (Fig. 1.2) is markedly in contrast with the systems previously described. *Vibrio harveyi* contains an interesting blend of Gram-negative and -positive quorum sensing mechanisms [55]. Two parallel systems converge to regulate *luxCDABE*, the luciferase structural operon. System 1, consisting of autoinducer-1 (AI-1) and sensor 1 (LuxN), is involved in the intraspecies quorum sensing. System 2, consisting of AI-2 and sensor 2 (LuxPQ), is used for interspecies cell-to-cell communication [56]. AI-1 [N-(4-hydroxybutyl)-L-homeserine] is a typical Gram-negative AHL, but it signals through 2CRS, like in Gram-positive bacteria. AI-2 does not resemble any other known signaling molecule and also signals through 2CRS. AI-2 is a furanosyl borate diester synthesized from SAM, an essential cofactor for DNA, RNA and protein synthesis, through at least three enzyme steps. Two enzymes, MetF (methylene tetrahydrofolate reductase) and MetE (cobalamin-independent methionine synthase), which are located upstream of LuxS, seem to be indispensable in the generation of methionine, which is part of the activated methyl cycle (AMC) [57]. Consumption of SAM, as a methyl donor, produces S-adenosylhomocysteine (SAH), which is hydrolyzed by nucleosidase Pfs to yield adenine and S-ribosylhomocysteine (SRH). LuxS catalyzes the cleavage of SRH to 4,5 dihydroxy-2,3-pentanedione (DPD) and homocysteine. DPD forms a cyclic molecule and undergoes further rearrangements to yield AI-2. Overall, the proposed AI-2 structure contains two fused five-membered rings, which are stabilized within the LuxP binding site through numerous polar interactions. Several candidates were considered for atom bridging the diester, with boron being the most suitable [58]. Another AI-2, which was characterized, is that of Salmonella Enteritidis [59]. The biosynthetic route to synthesize DPD is identical in *E. coli*, *V. cholerae*, *E. faecalis*, and *S. aureus* [60]. The regulatory network of *E. coli* is comprised of a transporter complex, LsrABCD, its repressor LsrR, and a cognate signal kinase LsrK [61]. Although the furanosyl borate diester is the most common form of AI-2, other furanone derivatives may assume the function of signaling molecules. For instance, two 2(5H)-furanones were released by *Lactobacillus helveticus* exposed to oxidative and heat stresses [62]. The exposure of this bacterium to 5-ehtyl-3-hydroxy-4-methyl-2(5H)-furanone also induced morphological changes.

DNA database analysis revealed that highly conserved homologues of *luxS* are present in over 30 species of both Gram-negative and -positive bacteria, including but not limited to *E. coli*, Salmonella Enteritidis, *B. subtilis*, *Campylobacter jejuni*, *E. faecalis*, *Streptococcus pyogenes*, *S. aureus* and *Clostridium perfringens* [63]. Currently, AI-2 is considered as the bacterial Esperanto, which could be used for interspecies communication [59]. The phylogenetic tree of the LuxS protein was reconstructed [64]. The Firmicutes forms a clad separated from Proteobacteria.

Within Proteobacteria, each division (α, β, and ϵ) is distinct. Within the monophyletic group Firmicutes, there are four sequences. The most surprising is the position of the LuxS of *Helicobacter pylori* as a sister group to *Staphylococcus*, *Listeria* and *Bacillus* species, and distant from *C. jejuni*. This suggests that *H. pylori* acquired the *luxS* gene from Firmicutes. *Bifidobacterium longum* was found to be a sister group to *L. plantarum*, *Lactococcus lactis*, and several *Streptococcus*. *B. longum* colonizes the human gastrointestinal tract where it is considered an important commensal. *Lactobacillus plantarum* and *Lc. lactis* are also known to be natural inhabitants of the human gastrointestinal tract. Thus, the proximity of the habitat of these species might explain the possibility of gene transfers among them. It is assumed that all LuxS-containing bacteria synthesize the DPD precursor. Probably, LuxS-containing bacteria release DPD or specific furanones, and then each recipient species acts on the precursor to generate a specific AI-2 signal. Several characteristics of the AI-2 biosynthetic pathway suggest that AI-2 harbors information on cell number, growth phase, and prosperity of the bacterial population. First, the extracellular accumulation of AI-2 is proportional to cell number. Second, one DPD molecule and, possibly, one AI-2 molecule, is synthesized every time that SAM is used as the methyl donor. This intimate link between SAM metabolism and AI-2 synthesis makes AI-2 an excellent device to measure the metabolic potential of the cell population. Third, the detoxification of a lethal intermediate (SAH) via Pfs ensures that the bacterial population will continue to grow (SAM is used) and to provide a substrate for the synthesis of AI-2. Conversely, if the population does not grow (SAM is not used) the synthesis of AI-2 will slow down [65].

1.5.1   The LuxS Paradigm

Despite the numerous studies on the role of LuxS in quorum sensing and in spite of the fact that the *luxS* gene is present in several Gram-negative and -positive bacteria, the use of the LuxS/AI-2 quorum sensing system is definitively demonstrated for only a few species [60, 66, 67]. The major difficulties concern the discrimination between the two potential roles of LuxS: as an enzyme involved in the AMC pathway or as an enzyme responsible for the synthesis of AI-2. Overall, the inactivation of the *luxS* gene should prevent either the AMC recycling intermediates to homocysteine or the synthesis of AI-2. Nevertheless, metabolomic and transcriptional analyses of a *luxS* mutant of *Lactobacillus reuteri* revealed the metabolic rather than the quorum sensing role of LuxS [68]. *L. reuteri* is an autochthonous inhabitant of the rodent forestomach, where it adheres to the nonsecretory epithelium and forms a biofilm. Microarray comparison of the gene expression between *L. reuteri* wild type and its *luxS* mutant revealed an altered transcription of genes encoding proteins, which were associated with cysteine biosynthesis/oxidative stress response, urease activity, and sortase-dependent proteins. Metabolomic analysis showed that the *luxS* mutation affected the levels of fermentation end products, fatty acids and amino acids. Cell density dependent changes of the gene transcription were not detected, thus hypothesizing that AI-2 was unlikely to be involved in gene regulation mediated by quorum sensing. The review "Bacterial cell to cell communication: sorry, can't talk now – gone to lunch" by Winzer et al. [60] furnishes some fundamental elements to discriminate between the dual role of LuxS. As the SAH degradation pathway via LuxS is present in many bacterial phylogenetic groups, its central role in cell metabolism is unquestionable. A comparison of bacterial genomes for genes, which are responsible for synthesis of AI-2, detoxification of SAH, and a signaling cascade to detect AI-2, revealed that LuxS is necessary to synthesize AI-2 but not to transduce the AI-2 signal. Theoretically, a microorganism may not have the capacity to synthesize AI-2 but it may detect coexisting or competing bacterial species by sensing the environmental concentration of AI-2. For instance, this is the case for *P. aeruginosa* [69]. Potential AI-2 receptors, such as those of *V. harveyi* (LuxPQ) and Salmonella Enteritidis (Lsr ABC-transporter), should be searched for also in other bacteria. The functionality of AI-2 receptors should be investigated in more detail, using approaches such as the growth of wild type on AI-2 depleted media or the growth of the *luxS* mutant on media complemented with chemically or biologically synthesized AI-2 [66]. Further elucidations are certainly needed, especially in those cases where an AI-2-mediated mechanism was claimed without previously detecting receptors in the bacterial genome sequences.

1.6 *LuxS/autoinducer-3*

The AI-3 molecule was casually decoded as a signal. This occurred during investigation of the quorum sensing mechanism, which regulates the expression of virulence genes of *E. coli* O157:H7. Initially, the expression of virulence and motility genes was ascribed to the signaling molecule AI-2. Further studies, which used purified and in vitro synthesized AI-2, showed that such genes were under the control of another autoinducer, AI-3 [70]. Differences between these two molecules were highlighted through biochemical analyses. In particular, AI-2 does not bind to C_{18} columns, whereas AI-3 does and it is eluted with methanol only. Electrospray mass spectrometry revealed that AI-3 is an aromatic aminated signal. Nevertheless, its complete structure is still unknown. Because of its low degree of hydrophobicity, it should not be able to cross the cell membrane. While the transcriptional assay for AI-2 is based on the induction of bioluminescence in *V. harveyi*, AI-3 does not show activity under this assay. AI-3 activates the transcription of the virulence genes of *E. coli* O157:H7, whereas AI-2 does not have this effect. As shown in Fig. 1.1, the synthesis of AI-2 depends on the LuxS enzyme. This is not the case for AI-3. Nevertheless, the *luxS* mutation leads to a decreased synthesis of AI-3. Further investigations suggested AI-3 may be an interkingdom signal as it cross-talks with the mammalian hormones adrenaline and norepinephrine to coordinate the interaction between host and bacteria [70]. This mechanism is described in more detail in Chap. 4.

References

1. Parsek MR, Greenberg EP (2005) Sociomicrobiology: the connections between quorum sensing and biofilms. Trends Microbiol 13:27–33
2. Duan K, Sibley CD, Davidson CJ, Surette MG (2009) Chemical interactions between organisms in microbial communities. In: Collin M, Schuch R (eds) Bacterial sensing and signaling, vol 16. Karger AG, Berlin, pp 1–17
3. Fuqua C, Parsek MR, Greenberg EP (2001) Regulation of gene expression by cell-to-cell communication: acyl-homoserine lactone quorum sensing. Annu Rev Gen 35:439–468
4. Kaplan HB, Greenberg EP (1985) Diffusion of autoinducer is involved in regulation of the *vibrio fischeri* luminescence system. J Bacteriol 163:1210–1214
5. Miller MB, Bassler BL (2001) Quorum sensing in bacteria. Annu Rev Microbiol 55:165–199
6. Fast W, Tipton PA (2012) The enzymes of bacterial census and censorship. Trends Biochem Sci 37:7–14
7. Uroz S, Dessaux Y, Oger P (2009) Quorum sensing and quorum quenching: the yin and yang of bacterial communication. Chem Bio Chem 10:205–216
8. Watson WT, Minogue TD, Val DL, von Bodman SB, Churchill ME (2002) Structural basis and specificity of acyl-homoserine lactone signal production in bacterial quorum sensing. Mol Cell 9:685–694
9. Gould TA, Schweizer HP, Churchill ME (2004) Structure of the *Pseudomonas aeruginosa* acyl-homoserinelactone synthase LasI. Mol Microbiol 53:1135–1146
10. Chung J, Goo E, Yu S, Choi O, Lee J, Kim J, Kim H, Igarashi J, Suga H, Moon JS, Hwang I, Rhee S (2011) Small-molecule inhibitor binding to an N-acyl-homoserine lactone synthase. Proc Natl Acad Sci USA 96:4360–4365
11. Parsek MR, Val DL, Hanzelka BL, Cronan JE, Greenberg EP (1999) Acyl homoserine-lactone quorum sensing signal generation. Proc Natl Acad Sci USA 96:4360–4365
12. Brader G, Sjöblom S, Hyytiäinen H, Sims-Huopaniemi K, Palva ET (2005) Altering substrate chain length specificity of an acylhomoserine lactone synthase in bacterial communication. J Biol Chem 280:10403–10409
13. Schaefer AL, Greenberg EP, Oliver CM, Oda Y, Huang JJ, Bittan-Banin G, Peres CM, Schmidt S, Juhaszova K, Sufrin JR, Harwood CS (2008) A new class of homoserine lactone quorum sensing signals. Nature 454:595–600
14. Arevalo-Ferro C, Hentzer M, Reil G, Görg A, Kjelleberg S, Givskov M, Riedel K, Eberl L (2003) Identification of quorum sensing regulated proteins in the opportunistic pathogen *Pseudomonas aeruginosa* by proteomics. Environ Microbiol 5:1350–1369
15. de Kievit TR, Iglewski BH (2000) Bacterial quorum sensing in pathogenic relationships. Infect Immun 68:4839–4849
16. Rasko DA, Sperandio V (2010) Anti-virulence strategies to combat bacteria-mediated disease. Nat Rev Drug Discov 9:117–128
17. Mattmann ME, Blackwel HE (2010) Small molecules that modulate quorum sensing and control virulence in *Pseudomonas aeruginosa*. J Org Chem 75:6737–6746
18. Schuster M, Greenberg EP (2008) LuxR-type proteins in *Pseudomonas aeruginosa* quorum sensing. Distinct mechanism with global implications. In: Winans SC, Bassler BL (eds) Chemical communication among bacteria. ASM Press, Washington, DC, p 161
19. Schuster M, Lostroh CP, Ogi T, Greenberg EP (2003) Identification, timing, and signal specificity of *Pseudomonas aeruginosa* quorum-controlled genes: a transcriptome analysis. J Bacteriol 185:2066–2079
20. Chugani SA, Whiteley M, Lee KM, D'Argenio DA, Manoil C, Greenberg EP (2001) QscR, a modulator of quorum sensing signal synthesis and virulence in *Pseudomonas aeruginosa*. Proc Natl Acad Sci USA 98:2752–2757
21. Lee J-H, Lequette Y, Greenberg EP (2006) Activity of purified QscR, a *Pseudomonas aeruginosa* orphan quorum sensing transcription factor. Mol Microbiol 59:602–609

22. Minagawa S, Inami H, Kato T, Sawada S, Yasuki T, Miyairi S, Horikawa M, Okuda J, Gotoh N (2012) RND type efflux pump system MexAB-OprM of *Pseudomonas aeruginosa* selects bacterial languages, 3-oxo-acyl-homoserine lactones, for cell-to-cell communication. BMC Microbiol 12:70–82
23. Pirhonen M, Flego D, Heikinheimo R, Palva TE (1993) A small diffusible signal molecule is responsible for the global control of virulence and exoenzyme production in the plant pathogen *Erwinia carotovora*. EMBO J 12:2467–2476
24. Chatterjee A, Cui Y, Chakrabarty P, Chatterjee AK (2010) Regulation of motility in *Erwinia carotovora* subsp. *carotovora*: quorum sensing signal controls FlhDC, the global regulator of flagellar and exoprotein genes, by modulating the production of RsmA, an RNA-binding protein. Mol Plant Microbe Interact 23:1316–1323
25. McGowan S, Sebaihia M, Jones S, Yu B, Bainton NJ, Chan PF, Bycroft BW, Stewart GSAB, Salmond GPC, Williams P (1995) Carbapenem antibiotic production in *Erwinia carotovora* is regulated by CarR, a homologue of the LuxR transcriptional activator. Microbiology 141:541–550
26. Medina-Martınez MS, Uyttendaele M, Meireman S, Debevere J (2007) Relevance of N-acyl-L-homoserine lactone production by *Yersinia enterocolitica* in fresh foods. J Appl Microbiol 102:1150–1158
27. Jacobi CA, Bach A, Eberl L, Steidle A, Heesemann J (2003) Detection of N-(3-oxohexanoyl)-L-homoserine lactone in mice infected with *Yersinia enterocolitica* serotype O8. Infect Immun 71:6624–6626
28. Ahmer BMM (2004) Cell-to-cell signalling in *Escherichia coli* and *Salmonella enterica*. Mol Microbiol 52:933–945
29. Michiels J, Dirix G, Vanderleyden J, Xi C (2001) Processing and export of peptide pheromones and bacteriocins in Gram-negative bacteria. Trends Microbiol 9:164–168
30. Flavier AB, Clough SJ, Schell MA, Denny TP (1997) Identification of 3-hydroxypalmitic acid methyl ester as a novel autoregulator controlling virulence in *Raistonia solanacearum*. Mol Microbiol 26:251–259
31. Holden MT, Ram Chhabra S, de Nys R, Stead P, Bainton NJ, Hill PJ, Manefield M, Kumar N, Labatte M, England D, Rice S, Givskov M, Salmond GPC, Stewart GSAB, Bycroft BW, Kjelleberg S, Williams P (1999) Quorum sensing cross talk: isolation and chemical characterization of cyclic dipeptides from *Pseudomonas aeruginosa* and other Gram-negative bacteria. Mol Microbiol 33:1254–1266
32. Nakayama J, Cao Y, Horii T, Sakuda S, Akkermans ADL, De Vos WM (2001) Gelatinase biosynthesis–activating pheromone: a peptide lactone that mediates a quorum sensing in *Enterococcus faecalis*. Mol Microbiol 41:145–154
33. Hellingwerf KJ, Crielaard WC, Teixeira J, de Mattos M, Hoff WD, Kort R, Verhamme DT, Avignone-Rossa C (1998) Current topics in signal transduction in bacteria. Ant van Leeuw 74:211–227
34. Quadri LEN, Kleerebezem M, Kuipers OP, De Vos WM, Roy KL, Vederas JC, Stiles ME (1997) Characterization of a locus from *Carnobacterium piscicola* LV17B involved in bacteriocin production and immunity: evidence for global inducer-mediated transcriptional regulation. J Bacteriol 179:6163–6171
35. Brurberg MB, Nes IF, Eijsink VGH (1997) Pheromone-induced production of antimicrobial peptides in *Lactobacillus*. Mol Microbiol 26:347–360
36. O'Keeffe T, Hill C, Ross RP (1999) Characterization and heterologous expression of the genes encoding enterocin A production, immunity, and regulation in *Enterococcus faecium* DPC1146. Appl Environ Microbiol 65:1506–1515
37. Clewell DB (1993) Bacterial sex pheromone-induced plasmid transfer. Cell 3:9–12
38. Morrison DA (1997) Streptococcal competence for genetic transformation: regulation by peptide pheromones. Microb Drug Resist Mech Epidemiol Dis 3:27–37
39. Magnuson R, Solomon J, Grossman AD (1994) Biochemical and genetic characterization of a competence pheromone from *B. subtilis*. Cell 77:207–216

40. Rudner DZ, LeDeaux JR, Ireton K, Grossman AD (1991) The *spo0K* locus of *Bacillus subtilis* is homologous to the oligopeptide permease locus and is required for sporulation and competence. J Bacteriol 173:1388–1398
41. Novick RP, Projan SJ, Kornblum J, Ross HF, Ji G, Kreiswirth B, Vandenesch F, Moghazeh S (1995) The *agr* P2 operon: an autocatalytic sensory transduction system in *Staphilococcus aureus*. Mol Gen Genet 248:446–458
42. Hoch JA (2000) Two-component and phosphorelay signal transduction. Curr Opin Microbiol 3:165–170
43. Geisinger E, Novik RP (2008) Signal integration and virulence gene regulation in *Staphylococcus aureus*. In: Winans SC, Bassler BL (eds) Chemical communication among bacteria. ASM Press, Washington, DC, p 161
44. Cheung AL, Coomey JM, Butler CA, Projan SJ, Fischetti VA (1992) Regulation of exoprotein expression in *Staphylococcus aureus* by a locus (*sar*) distinct from *agr*. Proc Natl Acad Sci USA 1992(89):6462–6466
45. Tegmark K, Karlsson A, Arvidson S (2000) Identification and characterization of SarH1, a new global regulator of virulence gene expression in *Staphylococcus aureus*. Mol Microbiol 37:398–409
46. McNamara PJ, Milligan-Monroe KC, Khalili S, Proctor RA (2000) Identification, cloning, and initial characterization of *rot*, a locus encoding a regulator of virulence factor expression in *Staphylococcus aureus*. J Bacteriol 182:3197–3203
47. Ji G, Beavis RC, Novick RP (1997) Bacterial interference caused by autoinducing peptide variants. Science 276:2027–2030
48. Lina G, Jarraud S, Ji G, Greenland T, Pedraza A, Etienne J, Novick RP, Vandenesch F (1998) Transmembrane topology and histidine protein kinase activity of AgrC, the *agr* signal receptor in *Staphylococcus aureus*. Mol Microbiol 28:655–662
49. Yarwood JM, Schlievert PM (2003) Quorum sensing in *Staphylococcus* infection. J Clin Invest 112:1620–1625
50. Qin X, Singh KV, Welnstock GM, Murray BE (2001) Characterization of fsr, a regulator controlling expression of gelatinase and serine protease in *Enterococcus faecalis* OG1RF. J Bacteriol 183:3372–3382
51. Wolfgang H, Shepard BD, Gilmore MS (2002) Two-component regulator of *Enterococcus faecalis* cytolysin responds to quorum sensing autoinduction. Nature 415:84–87
52. Sturme MHJ, Nakayama J, Molenaar D, Murakami Y, Kunugi R, Fujii T, Vaughan EE, Kleerebezem M, de Vos W (2005) An *agr*-like two-component regulatory system in *Lactobacillus plantarum* is involved in production of a novel cyclic peptide and regulation of adherence. J Bacteriol 187:5224–5235
53. Comella N, Grossman A (2005) Conservation of genes and processes controlled by the quorum response in bacteria: characterization of genes controlled by the quorum sensing transcription factor ComA in *Bacillus subtilis*. Mol Microbiol 57:1159–1174
54. Takano E, Chakraburtty R, Nihira T, Yamada Y, Bibb MJ (2001) A complex role for the gamma-butyrolactone SCB1 in regulating antibiotic production in *Streptomices coelicolor* A3(2). Mol Microbiol 41:1015–1028
55. Bassler BL, Wright M, Silverman MR (1994) Multiple signalling systems controlling gene expression of luminescence in *Vibrio harveyi*: sequence and function of genes encoding a second sensory pathway. Mol Microbiol 13:273–286
56. Gobbetti M, De Angelis M, Di Cagno R, Minervini F, Limitone A (2007) Cell-cell communication in food related bacteria. Int J Food Microbiol 120:34–45
57. Schauder S, Penna L, Ritton A, Manin C, Parker F, Renauld-Mongénie G (2005) Proteomics analysis by two-dimensional differential gel electrophoresis reveals the lack of a broad response of *Neisseria meningitidis* to in vitro-produced AI-2. J Bacteriol 187:392–395
58. Chen X, Schauder S, Potler N, Van Dorsselaer A, Pelczer I, Bassler BL, Hughson FM (2002) Structural identification of a bacterial quorum sensing signal containing boron. Nature 115:545–549

59. De Keersmaecker SCJ, Sonck K, Vanderleyden J (2006) Let LuxS speak up in AI-2 signaling. Trends Microbiol 14:114–119
60. Winzer K, Hardie KR, Williams P (2002) Bacterial cell-to-cell communication: sorry, can't talk now – gone to lunch! Curr Opin Microbiol 5:216–222
61. Wang L, Li J, March JC, Valdes JJ, Bentley WE (2005) luxS-Dependent gene regulation in Escherichia coli K-12 revealed by genomic expression profiling. J Bacteriol 187:8350–8360
62. Ndagijimana M, Vallicelli M, Cocconcelli PS, Cappa F, Patrignani F, Lanciotti R, Guerzoni ME (2006) Two 2[5 H]-furanones as possile signalling molecules in Lactobacillus helveticus. Appl Environ Microbiol 72:6053–6061
63. Bassler BL, Miller MB (2006) Quorum sensing. http://141.150.157.117:8080/prokPUB/chaphtm/320/COMPLETE.htm
64. Lerat E, Moran NA (2004) The evolutionary history of quorum sensing in bacteria. Mol Microbiol Evol 21:903–913
65. Xavier KB, Bassler BL (2003) LuxS quorum sensing: more than just a numbers game. Curr Opin Microbiol 6:191–197
66. Rezzonico F, Duffy B (2008) Lack of genomic evidence of AI-2 receptors suggests a non-quorum sensing role for luxS in most bacteria. BMC Microbiol 8:154–173
67. Vendeville A, Winzer K, Heurlier K, Tang CM, Hardie KR (2005) Making 'sense' of metabolism: autoinducer-2, LuxS, and pathogenic bacteria. Nat Rev Microbiol 3:383–396
68. Wilson CM, Aggio RBM, O'Toole PW, Villas-Boas S, Tannock GW (2012) Transcriptional and metabolomic consequences of luxS inactivation reveal a metabolic rather than quorum sensing role for LuxS in Lactobacillus reuteri 100–23. J Bacteriol 194:1743–1746
69. Sun J, Daniel R, Wagner-Döbler I, An-Ping Z (2004) Is autoinducer-2 a universal signal for interspecies communication: a comparative genomic and phylogenetic analysis of the synthesis and signal transduction pathways. BMC Evol Biol 4:36–47
70. Sperandio V, Torres AG, Jarvis B, Nataro JP, Kaper JB (2003) Bacteriahost communication: the language of hormones. Proc Natl Acad Sci USA 100:8951–8956
71. Winans SC, Bassler BL (2002) Mob Psychology. J Bacteriol 184:873–883

Chapter 2
The Phenotypes

2.1 Introduction

Once the main bacterial languages were partly decoded, the main efforts were consequently addressed to understanding the relevant phenotypes, which are coordinated in a cell density-dependent manner. N-acyl-L-homoserine lactones (AHL), autoinducing peptide (AIP, or peptide pheromone), autoinducer-2 (AI-2), through the activity of LuxS, and the new autoinducer-3 (AI-3), are all bacterial signals that may induce a large number of phenotypes. Competence, virulence, synthesis of toxins and exopolysaccharides (EPS), biofilm formation, and production of secondary metabolites are some examples of the above phenotypes, which are directly or indirectly under the control of quorum sensing circuits. Several of these phenotypic tracts and the related mechanisms of control may be of marked interest in relation to foods, either in terms of sensory and nutritional quality or considering foods themselves as vehicles of pathogenic bacteria.

Although some results are still controversial, the main findings concerning some of the above phenotypes are described in the following.

2.2 Virulence

Virulence genes encode proteins whose functions are essential to effectively establish a bacterial infection in the host organism. In many Gram-negative and -positive bacteria, the expression of some virulence factors is regulated by quorum sensing.

The language of *Pseudomonas aeruginosa* comprises N-acyl-L-homoserine lactones (AHL) and 4-quinolone quorum sensing signals (see Sect. 1.3). Two systems (*las* and *rhl*) encode the transcriptions of the regulatory proteins (LasR or RhlR) and autoinducer synthases (LasI or RhlI), and regulate the surface-associated or secreted virulence factors [1, 2]. As shown using animal models, when the mutation of *las*

M. Gobbetti and R. Di Cagno, *Bacterial Communication in Foods*,
SpringerBriefs in Food, Health, and Nutrition, DOI 10.1007/978-1-4614-5656-8_2,
© Marco Gobbetti and Raffaella Di Cagno 2013

and/or *rhl* occurred, the pathogenesis of *P. aeruginosa* decreased [3]. Comparisons between the secretome of *P. aeruginosa* wild type and one or more *las* and *rhl* mutants showed lower levels of proteins were released in the latter two [4]. This suggested that the lack of *las* or *rhl* severely disrupts the secretion of proteins and/ or the expression of abundant extracellular constituents. Unknown quorum sensing regulated proteins such as aminopeptidase PA2939, endoproteinase PrpL and unique hypothetical protein PA0572 were identified. The *las* mutant did not express the major isoforms of the aminopeptidase PA2939, which contains a putative signal [5]. Under starvation conditions, PA2939 generates free amino acids from short peptides. The *rhl* mutant did not express the endoproteinase PrpL, which has the capacity to cleave lactoferrin, transferrin, elastin, and casein. The azurin precursor, chitin-binding protein (CbpD), and the hypothetical protein PA4944 were only found in *P. aeruginosa* wild type. CbpD has adhesion-like properties and is protected from proteolysis by elastase, when it is bound to chitin [6]. The hypothetical protein PA4944 has high sequence similarity to host factor I, a RNA-binding protein that regulates the synthesis of enterotoxin in *Yersinia enterocolitica* [7] and several virulence factors in *Brucella abortus* [8]. Another protein considered to belong to the family of quorum sensing regulators is PA4944. Two partner secretion exo-proteins (PA0041 and PA4625), and quorum sensing regulated extracellular proteins (LasB elastase, LasA protease and aprA alkaline metalloproteinase) were found at the highest levels in the culture supernatants of quorum sensing mutants. This suggested that quorum sensing might also negatively control the expression of some functional genes for virulence.

The regulation of virulence factors from soft-rotting plant pathogen *Erwinia carotovora* occurs via AHL signaling (see Sect. 1.3). In addition to brute force virulence factors, *E. carotovora* also produces extracellular enzymes as secondary metabolites and multiple subtle virulence factors [9]. Regulation of secondary metabolite systems AB (RsmAB) were identified in *E. carotovora* subspp. *carotovora* and *atroseptica*. A mutant defective of *rsmA* exhibited the over production of extracellular enzymes and caused disease bypassing the quorum sensing system. The Rsm system of *E. carotovora* appears to function similarly to the Csr system of *Escherichia coli*. RsmA represses extracellular enzymes by promoting transcript degradation. On the contrary, RsmB is thought to bind to RsmA and to prevent it from binding to its target transcripts, thus indirectly mediating the activation of extracellular enzymes. The quorum sensing locus of *E. carotovora* controls the Rsm system through *rsmA* and, conversely, the Rsm system affects the quorum sensing machinery by modulating the expression of *expI* (homologous to *luxI*) and the consequent synthesis of AHL. As many virulence factors are under the control of quorum sensing in *E. carotovora*, this suggests that the role of quorum sensing during infection is more complicated than simply orchestration by AHL. Examples of virulence factors are Svx, a necrosis-inducing protein, and harpin HrpN, an extracellular glycine-rich protein that elicits the hypersensitive reaction. Several other secreted proteins (e.g., ECA0852 and ECA2220) were quorum sensing dependent, which makes them good candidates for novel virulence factors.

Virulence factors are considered to be terminal virulence determinants to which the plant is exposed. Nevertheless, processes other than gene expression have to be fulfilled by AHL or other secreted virulence factors to interact with the plant. For instance, the relevant virulon has to be secreted from the bacterial cell. Hence the system of protein secretion is an accessory virulence determinant, which is also subjected to quorum sensing regulation. Lip type I of *Serratia liquefaciens* and the Xcp type II of *P. aeruginosa* were identified as secretion systems, which are quorum sensing dependent.

Burkholdeira cenocepacia is a common inhabitant of soil, water, and plant surfaces where it may cause diseases such as the soft rot of onion bulbs. The bacterium is also an opportunistic pathogen, especially, in patients that are affected by cystic fibrosis. The quorum sensing system of *B. cenocepacia* uses the AHL synthase CepI, which directs the synthesis of N-octanoylhomoserine lactones [10, 11], and CepR, which activates or represses the transcription of target genes. The *cep* system regulates biofilm formation, swarming motility, synthesis of extracellular proteolytic and chitinolytic enzymes, and represses the synthesis of the siderophore ornibactin [10, 12, 13]. The proteomes of *B. cepacia* H111 wild type and that of the *cep* mutant were compared [14]. Fifty of the ca. 1,000 proteins detected were differentially expressed. Addition of AHL molecules to the culture medium restored the protein profile of the *cep* mutant. About 5 % of the *B. cepacia* proteome was down-regulated and 1% up-regulated in the *cep* mutant. A number of apparently unrelated functions seemed to be *cep* regulated, including the activity of the peroxidase RSC0754 and superoxide dismutase (SodB). The synthesis of SodB by *P. aeruginosa* increased during biofilm formation [15, 16]. This suggested that the *cep* quorum sensing system provides a regulatory link between surface colonization and development of resistance against oxidative stress.

Vibrio vulnificus is a Gram-negative human pathogen, in some cases conveyed by foods. A number of factors are implicated in its virulence and pathogenesis: capsular polysaccharide, lipopolysaccharide (LPS), elastase, cytolysin, metalloprotease, siderophores, and phospholipase [17, 18]. The quorum sensing of *V. vulnificus* is controlled through a hierarchical circuit via *luxS* and *smcR* (homologous to *luxR*). The proteome profile of the *luxS-smcR* double mutant was compared to that of the wild type [19]. Some proteins were repressed by double mutation. They included Zn-dependent protease (VVP), which is responsible for skin lesions [20], periplasmic ABC-type Fe^{3+} transport system and deoxyribose-phosphate aldolase (DERA), which determine the adaptation to starvation and/or the deoxynucleoside catabolism [21], and phosphomannomutase (PMM), which is responsible for the biosynthesis of EPS and LPS.

The global protein expression was compared between the pathogenic wild type *Escherichia coli* O157:H7 and its isogenic *luxS* mutant, and between the *luxS* mutant and *luxS* mutant supplemented with AI-2 [22]; 11 and 18 proteins were differentially expressed, respectively. Both comparisons showed differential expression of the tryptophan repressor binding protein (WrbA), phosphoglycerate mutase (GpmA), and putative protein YbbN. The up-regulation of the FliC protein, which is responsible for flagellar synthesis and motility, was only found in the wild type. The addition of AI-2 did not influence the synthesis of FliC by the *luxS* mutant. This suggested that signaling molecules other than AI-2 are involved in flagellar synthesis and motility. Overall, flagellar synthesis and motility are strictly related to virulence phenotypes. A comparison was also made of *E. coli* O157:H7 and its *luxS* mutant under the probiotic effect (inhibition of AI-2 like activity) of the cell extract of *Lactobacillus acidophilus* A4. Five proteins (NifU, PapC, FlgI, MdaB, and DsbA), which are responsible for pathogenesis, were up-regulated in the presence of AI-2 activity (wild type) or down-regulated in the *luxS* mutant and wild type subjected to the probiotic effect. These findings showed the relationship between AI-2 and the virulence of *E. coli* O157:H7 as well as the potential role of *L. acidophilus* as a quenching agent.

PlcR is the major virulence regulator of the *Bacillus cereus* group, which includes species that very often contaminate vegetable foods [23]. In addition to *B. cereus* sensu stricto, an opportunistic pathogen that causes gastroenteritis, pneumonia and endophthalmitis, this group includes *Bacillus thuringiensis,* an entomopathogenic bacterium used to produce biopesticides, and *Bacillus anthracis,* the causative agent of anthrax [24, 25]. The activity of PlcR depends on PapR, a secreted signaling peptide re-imported into the bacterial cell through the Opp transport system [26]. When high bacterial density is reached, the intracellular concentration of PapR increases, which promotes its interaction with PlcR. Then, the PapR: PlcR complex binds to its DNA recognition site, the palindromic PlcR box, and triggers a positive feedback loop that up-regulates the expression of *plcR*, *papR*, and various virulence factors [26]. The molecular basis for transcriptional control by PapR: PlcR is still unknown.

Clostridium perfringens uses AI-2/LuxS to regulate the toxin production [27]. The timing of toxin production is critical for the virulence of this species, which occurs at the mid-late exponential phase of growth. This maximum synthesis of the toxin coincides with the maximum synthesis of AI-2. Compared to wild type, *C. perfringens luxS* mutants have reduced toxin transcription at the mid-late exponential phase of growth, whereas levels of the toxin mRNA were similar in the stationary phase of growth.

2.3   Biofilm Formation

Bacteria develop a biofilm on a number of different surfaces, such as natural aquatic and soil environments, living tissues, vegetables and fruits, medical devices or industrial or potable water piping systems [28, 29]. Biofilm formation is a prerequisite for the existence and survival of microbial aggregates [29, 30]. EPS are the main components of biofilms, even though the type of EPS varies according to the status of bacterial growth and the substrate for microbial metabolism. As almost all bacterial species that form biofilms may synthesize and degrade EPS, these latter are considered tools for communication. Bacteria living attached to surfaces and enveloped within biofilms substantially differ from planktonic cells [31]. EPS provide shelter to bacteria, block harmful agents, and trap nutrients from the environment thereby increasing the local concentration.

Formation of a biofilm is a complex process, which is regulated at different stages via diverse mechanisms [32, 33]. The most-studied regulatory mechanism is quorum sensing [28, 32–35]. At a given population density, the genes responsible for biofilm differentiation and maturation are activated [28, 33]. During growth within a biofilm, cells are in close contact with their neighbors and this promotes communication [36]. A few examples were described from mixed cultures during food and beverage fermentation [37]. The synthesis of the capsular kefiran (a type of EPS) promoted physical contact between *Lactobacillus kefiranofaciens* and *Saccharomyces cerevisiae*, the usual natural starters for making kefir. It was postulated that bacteria and yeasts benefit from the synthesis of kefiran, which promote interactions within the kefir grains, where the exchange of growth factors is facilitated. Under biofilm conditions, the synthesis and activity of bacteriocins is more efficient. Killing sensitive strains within a delimited zone, around the bacteriocin-producing strain, favors a more efficient increase of available nutrients than that found under broth culture conditions [38].

The role of quorum sensing in biofilm formation cannot be described in general terms but it varies depending on the bacterial species [39]. Quorum sensing is essential for adhesion, biofilm formation, and virulence of *P. aeruginosa* [33]. Mutants of *P. aeruginosa* that did not synthesize quorum sensing signals formed thinner biofilms than the wild type. Mutation of the *lasI* gene also resulted in an abnormal and undifferentiated biofilm formation [40]. The link between quorum sensing and the biofilm seemed to be mediated via the synthesis of EPS, with the unknown protein PA1324 having the role of binding and transporting EPS during biofilm formation [41]. Another important biofilm component is the polysaccharide intercellular adhesin (PIA), which mediates cell-to-cell adhesion [42]. Glucose is required to synthesize PIA [43], and uridine diphosphate-N-acetylglucosamine (UDP-GlcNAc) is the precursor of the polysaccharide matrix [44]. Indeed, the addition of glucose and UDP-GlcNAc in the culture medium stimulated the synthesis of PIA and the formation of a biofilm by *P. aeruginosa* [44]. N-acetylglucosamine is also the repeating unit within the heparin molecule, which stimulates the formation of a biofilm [45]. Heparin favors the adherence of *P. aeruginosa* to epithelial respiratory cells [46]. *Pseudomonas aeruginosa* also synthesizes alginate as the main biofilm component, which is made up of glucose, galactose, and pyruvate [47].

LuxS is required for biofilm formation on human gallstones by Salmonella Enteritidis [48]. Formation of a biofilm on gallstone surfaces should offer long-term protection against antimicrobial agents and high concentrations of bile. Salmonella Enteritidis senses the presence of bile as a signal. This induces the synthesis of bacterial surface organelles (e.g., fimbriae, flagella), which promote the formation of a biofilm. Flagella play a role in the secretion or synthesis of EPS as well as in the initial adherence and formation of micro-colonies. The comparison of the global protein expression between the wild type and *luxS* mutant of Salmonella Enteritidis showed the negative effect of LuxS on the synthesis of flagellin [22]. The proteome of Salmonella Enteritidis was studied under conditions that mimicked the in vivo infection [49]. Two-dimensional differential in gel electrophoresis (2-D DIGE) analysis showed that adaptation was mediated through up- and down-regulation of several proteins. In particular, the uptake of AI-2, and the expression of LsrF, LsrA, LsrB, and LsrR were up-regulated. LsrA and LsrB are part of the AI-2 uptake transporter. Once AI-2 is phosphorylated, it binds to the transcriptional repressor LsrR. As such, it alleviates the repression of the *lsr* operon and allows the increased transcription of the *lsr*-genes, which resulted in an increased internalization of AI-2 [50]. It is supposed that stream of AI-2 is executed by LsrF, LsrE, and LsrG [50]. The up-regulation of LsrA and LsrB was related to the pathogenesis of Salmonella Enteritidis via the activation of the transcriptional regulator PhoP. This is a part of the two-component regulatory system (2CRS), which senses the concentration of extracellular Mg^{2+} [51].

Biofilm formation and architecture, and cell fimbriae were significantly altered in *lsrR* and *lsrK* mutants (see Sect. 1.5) of *E. coli* [52]. While *H. pylori* secretes EPS during biofilm formation [53] and other enteric pathogens such as *Salmonella* also use carbohydrates extensively [54], the matrix surrounding the biofilm of *Campylobacter jejuni* remains to be defined. The genome of *C. jejuni* encodes a limited repertoire of regulatory elements, which include a relatively small number of 2CRS [seven histidine protein kinase (HPK) and 12 response regulators (RR)] [55]. The CprRS sensor kinase mutant of *C. jejuni* displayed an apparent growth defect, and formed an enhanced and accelerated biofilm [56]. Modifications were consistent with the modulation of essential metabolic genes, and up-regulation of stress-tolerance proteins and cell surface structures. Oxidative stress-tolerance proteins such as catalase (Kat), thioredoxin reductase (TrxR), and alkyl hydroperoxide reductase (Ahp) were up-regulated. The major outer membrane protein and flagellar filament protein FlaA were also up-regulated. Down-regulation was found for the orphan RR and LuxS. The diversity of the deregulated proteins suggested that CprRS controls various aspects of *C. jejuni*, and the hypothesis was that nutrient availability might influence the formation of a biofilm.

An *agr*-like 2CRS, which encodes a cyclic thiolactone autoinducing peptide (AIP, LamD558), was found in *Lactobacillus plantarum* WCFS1 (Fig. 2.1) [57] (see Sect. 1.4). LamD558 has a ring structure similar to that of AIP from the staphylococcal *agr* system and it is involved in the regulation of adherence. Complete *agrBDCA*-like systems were found only for pathogenic bacteria such as staphylococci [58], *Enterococcus faecalis* [59] and *Listeria monocytogenes* [60]. Similarly, the *lamBDCA* system of *L. plantarum* may play a role in commensal host-microbe interaction [61].

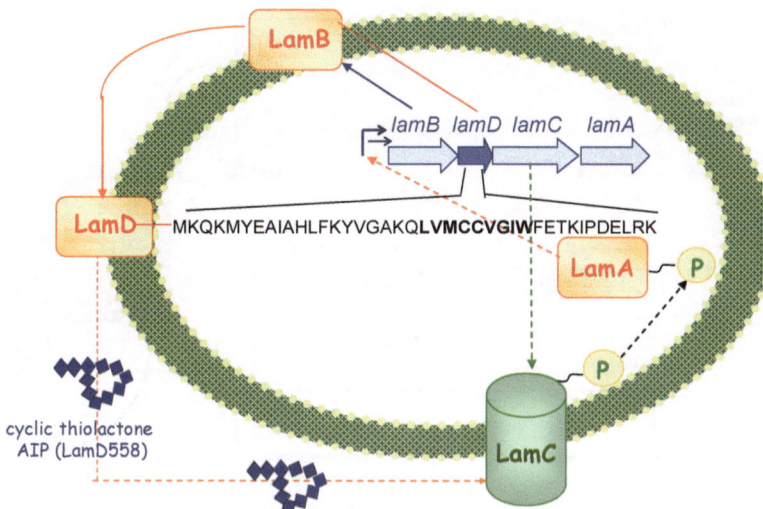

Fig. 2.1 Schematic representation of an agr-like two-component regulatory system (2CRS) found in *Lactobacillus plantarum* WCFS1. The *lam* quorum sensing system encodes the two-component histidine protein kinase LamC and response regulator LamA, an autoinducing pentapeptide (*AIP*) cyclic thiolactone derived from precursor peptide LamD and additionally LamB, a protein involved in processing and post-translational modification of LamD. The signal cyclic thiolactone pentapeptide with a ring structure was designated as LamD558. Amino acids of predicted AIP sequence is shown in bold type (Adapted from [22])

2.4 Bacteriocin Synthesis

Bacterial communities produce antimicrobial compounds to compete with other similar microorganisms. On the basis of biosynthetic mechanisms, bacteria produce two types of antimicrobial peptides: ribosomally synthesized peptides, or bacteriocins, which exhibit a relatively narrow range of antimicrobial activity, mainly inhibiting closely related bacteria that share the same ecological niche [62]; and nonribosomally synthesized peptides that show broader spectra of activities, inhibiting bacteria or fungi. On the basis of biochemical and genetic properties, bacteriocins are grouped into four classes (I–IV) [63]. Both class I and II bacteriocins are small (3–10 kDa), cationic, amphiphilic, and membrane-active peptides. Class I bacteriocins, or lantibiotics, contain the unusual amino acids lanthionine and methyllanthionine. On the contrary, class II bacteriocins do not contain these modified amino acids. They are subdivided into three classes: IIa, *Listeria*-active peptides with the consensus sequence -Y-G-N-G-V-X-C- near the N-terminus; IIb, two-peptide bacteriocins, in which both components are required for antimicrobial activity; and IIc, thiol-activated peptides that require reduced cysteine residues for activity. Class III bacteriocins are high molecular mass (>30 kDa), heat-labile proteins. Class IV bacteriocins are complex peptides containing lipid or carbohydrate moieties, which are essential for activity.

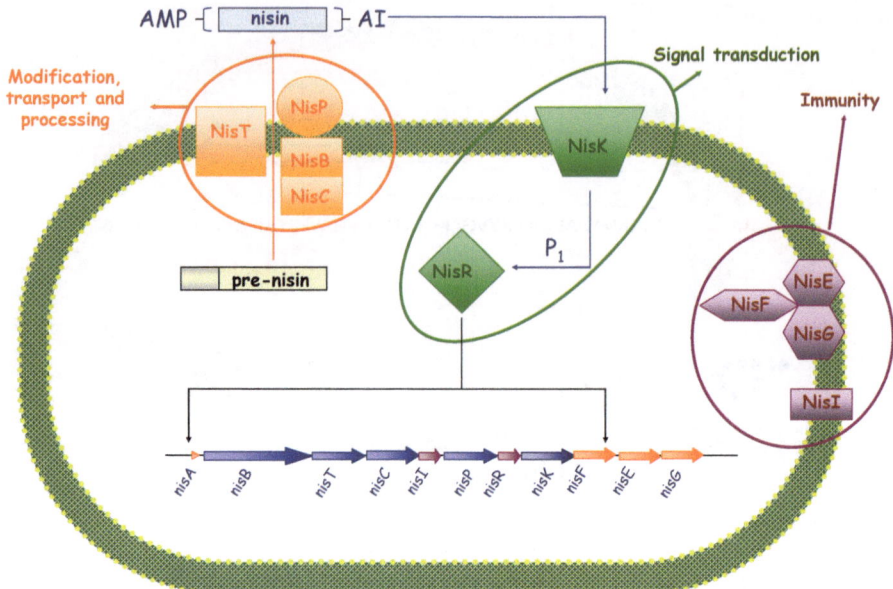

Fig. 2.2 Quorum sensing regulation of class I antimicrobial peptides (*AMP*) in lactic acid bacteria. *NisABTCIPRKFEG*, gene cluster encoding nisin; NisB and NisC, proteins involved in the intracellular post-translational modification reactions; NisT, putative transport protein of the ABC translocator family; NisP, extracellular protease for removing the leader peptide; AI, autoinducer; NisK, transmembrane-associated signal transducer; NisR, response regulator; NisF, NisE and NisG, ABC exporter system that generates immunity through active cell extrusion from the cell; and NisI, lipoprotein that contributes to producer immunity. For the quorum sensing mechanism see the text (Adapted from [87])

Many Gram-positive bacteria, especially lactic acid bacteria, secrete small antimicrobial peptides (AMP) or bacteriocins, which are regulated via quorum sensing mechanisms [64]. These compounds are of marked interest as natural food preservatives [65] and/or because they exert inhibitory activity against pathogens at the gastrointestinal level of humans and animals [66]. Nisin, which is synthesized by *Lactococcus lactis*, is the best known and most used lantibiotic [67]. Nisin is produced as the 57-residue precursor that contains the 23-residue N-terminal extension, called the leader peptide, which is absent in the mature molecule. The biosynthesis of nisin is encoded by the gene cluster *nisABTCIPRKFEG* [68]. Besides the structural, processing and producer-immunity genes, the cluster also contains elements of the 2CRS system, RR (*nisR*) and HPK (*nisK*), which are responsible for the regulation of nisin biosynthesis (Fig. 2.2). The synthesis of nisin starts at the early to mid logarithmic phase of growth and increases to the maximal level at the early stationary phase of growth, when the highest cell density is reached. Introduction of a 4 bp deletion on the structural *nisA* gene (Δ*nisA*) of *Lc. lactis* resulted not only in the loss of the capacity to synthesize nisin but also in the abolition of Δ*nisA* transcription. The transcription of Δ*nisA* was restored by the addition of sub-inhibitory levels of nisin [69]. Therefore, besides its function as AMP, nisin also acts as a secreted signal molecule that induces the transcription of the genes involved in its biosynthesis. The signal transduction is mediated via NisK and NisR. The lantibiotic subtilisin of *Bacillus subtilis* is subjected to a similar quorum sensing circuit, which contains genes encoding HPK (*spaK*) and RR (*spaR*) [70]. A dual mechanism regulates the expression of subtilin. First, the σ factor H allows the low level of expression of the 2CRS SpaR/SpaK [71]. Further, subtilin auto-induces the histidine kinase SpaK, which, in turn, phosphorylates the response regulator SpaR and up-regulates the transcription of subtilin and immunity genes [71]. A novel subtilin-like lantibiotic, termed entianin, was identified in *B. subtilis* [72]. Combining DNA and mass spectrometry (MS/MS) sequencing data, it was shown that entianin exhibits the primary sequence of subtilin, except for the amino acid exchanges between Leu6 and Val6, Ala15 and Leu15, and Leu24 and Ile24. It represents the third subtilin-like lantibiotic along with ericin [73]. Entianin is synthesized in succinylated or unsuccinylated forms. In the latter case, the antimicrobial activity is much higher. Succinylation seems to dramatically decrease the antimicrobial activity. This is probably due to the diminished interaction between lipid II and lantibiotic or to the hampered integration of the complex into the cytoplasmic membrane. On the contrary, auto-induction is not adversely affected by succinylation. The *etn* gene cluster, which is responsible for entianin biosynthesis, regulation and autoimmunity, showed a high degree of homology (ca. 93 %) with the *spa* gene cluster that is responsible for subtilin biosynthesis. On the basis of genome sequences, the 2CRS of *Streptococcus thermophilus*, which consists of response regulator (RR) 04 (2CRS04), displays high homology with SpaK/SpaR of *B. subtilis* and NisK/NisR of *Lc. lactis* [74]. The biological relevance of this general regulatory mechanism, which is quite common to the above bacterial species, was based on the following considerations: (1) it ensures that the environmental concentration of AMP rapidly reaches levels, which are efficient to kill competitors; (2) the rapid increase of the concentration of AMP prevents the development of immunity mechanisms into target cells; and (3) it protects the producing cells from the ineffective activity, which may occur when AMP diffuses away from the environment [68].

Class II bacteriocins are synthesized as precursor peptides that contain an N-terminal extension, which is removed during or shortly after secretion of the peptide. Pro-peptides share the common feature of having two glycine residues (Gly-Gly motif) that precede the cleavage site. The genetic characterization of several strains of *L. plantarum*, which were variously isolated from vegetables, fermented foods and human saliva, showed that the same determinants were responsible for bacteriocin biosynthesis and gene regulation. These strains synthesized bacteriocins belonging to the group of plantaricins and their *pln* loci is bi-faceted, one part being highly conserved and the other mosaic like. The *pln* loci encode class IIb (plantaricins EF, JK, NC8, and J51) or class IIc (pheromone peptide plantaricin A, plnA) bacteriocins, one conserved ABC-transporter dedicated to export peptides, with the so called double-glycine leader, and two divergent quorum sensing networks. Many bacteriocins from lactic acid bacteria are only synthesized in broth cultures. This occurs when specific inoculum size and growth conditions are achieved, and a dedicated three-component regulatory system (3CRS), involved in quorum sensing mechanisms, is switched on. On the contrary, a few other bacteriocins are phenotypically constitutive, and they are synthesized on solid but not in liquid media. Such divergence in biosynthesis is usually attributed to differences in the rate of diffusion. Compared to a liquid medium, cells growing on the agar surface are in closer contact with the secreted bacteriocins. The question about constitutive and regulated bacteriocins was highlighted constructing knockout mutants for regulatory operons [75]. It was revealed that the synthesis of bacteriocins is under the control of quorum sensing mechanisms both on solid and liquid media. During growth in a liquid medium, the synthesis of bacteriocins occurs only in the presence of an elevated inoculum size or if an external source of bacteriocin is added to the medium. This confirmed the auto-induction mechanism. Such a regulatory mechanism was also shown for the synthesis of carnobacteriocin A, B2 and BM1 by *Carnobacterium piscicola* [76], several different putative plantaricins (PlnJK, PlnEF and PlnN) by *L. plantarum* [77], and sakacin P by *Lactobacillus sakei* [78]. The phenotype (Bac⁺) was lost upon inoculation of an overnight culture into fresh culture medium at the level below the threshold of inoculum size (10^6–10^4 cfu/ml). The Bac⁻ phenotype persisted during subsequent cultivation but it was recovered by addition of cell-free Bac⁺ culture supernatant. Other environmental factors are, probably, responsible for the synthesis of phenotypically constitutive bacteriocins into solid media [79]. Whatever be the case, most of these bacteriocins are synthesized in those culture conditions, which better mimic the natural ecological niche of lactic acid bacteria (e.g., growth on a solid surface and presence of inducing microorganisms) [80]. This phenotype should be of importance in food fermentation, especially for vegetables (e.g., olive fermentation), where solid matrices represent enormous surfaces for bacteria to adhere via biofilm formation. Under these ecological conditions, bacteria may find suitable environmental parameters to synthesize bacteriocins. Selection of starter cultures of *L. plantarum* for vegetable fermentations should also consider these features.

2.4.1 The Regulatory Operons and Their Regulated Promoters

As stated above the synthesis of bacteriocins is regulated through a quorum sensing pathway via 3CRS. Usually, this regulation involves three proteins: the secreted peptide autoinducing pheromone (AIP), the membrane-located histidine protein kinase (HPK), and the response regulator (RR). The secreted pheromone serves as a tool for measuring the cell density of the producer strain. At a certain cell density, AIP reaches the critical threshold concentration and triggers a cascade of phosphorylation, which culminated with the phosphorylated RR. This latter binds to the promoters of the bacteriocin regulon and activates the genes for biosynthesis. The *pln* regulon of *L. plantarum* C11 was studied in detail. The regulatory operon *plnABCD* codes for an auto-regulatory circuit, which activates its own transcription as well as the transcription of another four operons at the *pln* locus [77]. *plnABCD*

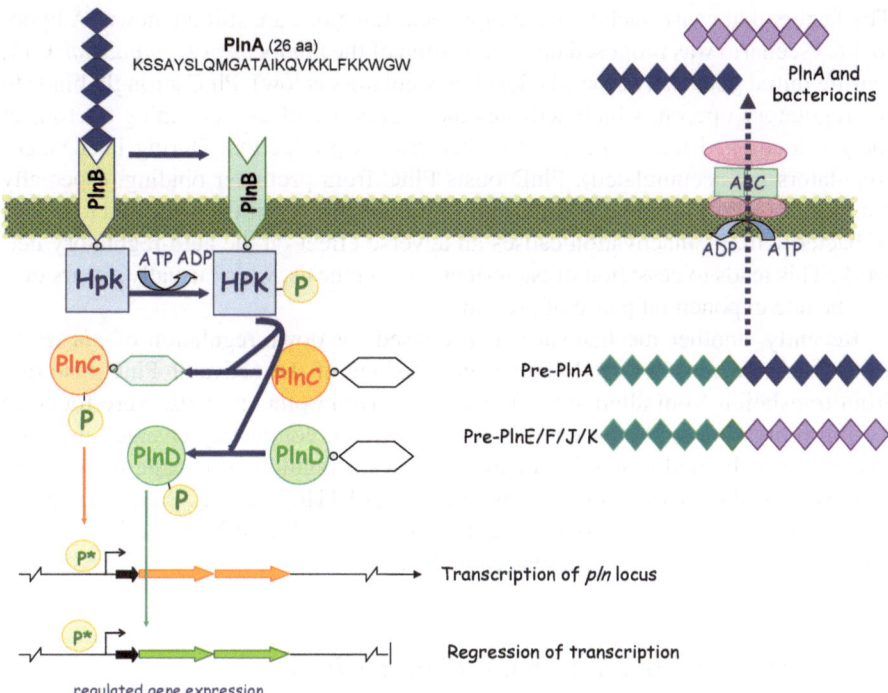

Fig. 2.3 Auto-regulatory network of the *pln* regulon in *Lactobacillus plantarum* C11. Binding of the inducing factor (*PlnA*) to the membrane domain of the histidine protein kinase PlnB leads to auto-phosphorylation of the cytoplasmic domain of PlnB and the subsequent transfer of the phosphoryl group to the gene regulators PlnC and PlnD. Phosphorylated regulators bind to regulated promoters to activate (by *PlnC*) or repress (by *PlnD*) expression of the genes involved in bacteriocin synthesis, including the auto-regulatory operon (*plnABCD*). All bacteriocins and the inducing peptide PlnA apply double-glycine leaders for export through a dedicated ABC transporter (Adapted from [81])

codes for plantaricin A (AIP), PlnB (homologous to HPK), and PlnC and PlnD (homologues to RR) [81] (Fig. 2.3). Almost the same regulatory network was found for *L. plantarum* NC8 and DC400, which were isolated from vegetables and Italian sourdoughs, respectively [81, 82]. Unlike the *pln* regulatory operon of *L. plantarum* C11, that of strain NC8 contains only three genes: *NC8-IF*, *NC8-HK*, and *NC8-plnD* that code for AIP, HPK, and RR, respectively. In general, the interactions between the peptide pheromones (plantaricin A or NC8-IF) and their cognate HPK (PlnB or NC8-HK) is specific and no cross-talk occurs between the same pheromones and noncognate HPK molecules. In vitro studies showed that both response regulators, PlnC and PlnD, bind as homo-dimers in a cooperative manner. When examined in a heterologous host (e.g., *L. sakei*), PlnC and PlnD act as positive regulators, the first being much stronger [81]. Nevertheless, when these regulators were individually overexpressed in the endogenous host (*L. plantarum* C11), they acted differently. PlnC activated, while PlnD repressed the biosynthesis of the bacteriocin [81].

The factors that cause such variable biological functions are still unknown. A hypothetical scenario was proposed upon activation of the *pln* locus of *L. plantarum* C11. During initial gene activation (the level of regulators is low), PlnC strongly binds to the regulatory operon, which activates the expression of the remaining operons at the *pln* locus and leads to a burst of bacteriocin production. During later stages (regulators are accumulated), PlnD ousts PlnC from promoter binding, especially from the transport promoter. As the encoded transport system is dedicated to export of bacteriocin, its inactivation causes an adverse effect on the auto-regulatory network. This leads to cessation of bacteriocin biosynthesis, which usually occurs during the late exponential phase of growth.

Recently, another mechanism was proposed for down-regulation of *plnABCD* from *L. plantarum* C11 [83]. Truncated versions of the activator PlnC resulting from translation from alternative start codons within plnC in cells, were found to exhibit repression of the bacteriocin regulon, which completely changed its functionality. It exhibited repression of the bacteriocin regulon. The same finding was observed for the bacteriocin systems of *L. sakei* LTH673 and *L. plantarum* NC8. This mode of repression may represent a common tool used by bacteria to down-regulate certain quorum sensing-based pathways.

2.4.2   The Peptide Pheromone Plantaricin A

Plantaricin A (PlnA) has a dual function in the plantaricin system. It works as an induction factor in gene regulation and as an antimicrobial peptide [84]. Plantaricin A was originally described as bacteriocin [85], and, in this context, it should be considered as belonging to class IIc: non-pediocin-like, one-peptide bacteriocin without post-translational modifications. The antimicrobial spectrum of PlnA is relatively narrow. It mainly comprises *Lactobacillus* species, for instance *Lactobacillus casei*, *L. sakei* and *Lactobacillus viridescens*, in addition to *L. plantarum* strains. Compared to other plantaricins (EF and JK), PlnA shows significantly lower activity, being 10–100-fold less potent [81]. Contrary to most of the bacteriocins, it lacks a dedicated immunity protein. These features suggest that PlnA is primarily an induction factor and that the antimicrobial activity is secondary, probably caused by the amphiphilic characteristics of its secondary structure.

From the structural point of view, PlnA is unstructured in aqueous solution, but it adopts an amphiphilic α-helix, from residue 12 to 21 (C-terminal part), when it comes in contact with negative charges into the membrane. The α-helix conformation is essential for pheromone and antimicrobial activities. Regarding the pheromone function, the α-helix facilitates the positioning of the N-terminal part of PlnA, which engages chiral interactions with the receptor PlnB. For antimicrobial activity, no chiral interactions take place and only the α-helix structure is sufficient to permeabilize sensitive cells [84]. Because of the necessity for the contact of PlnA with the membrane for it to act as a pheromone, it was suggested that the antimicrobial activity is a side effect, which is indirectly caused by the mode of action of the pheromone peptide.

As shown for other strains that populate other food ecosystems [86], multidimensional high-performance liquid chromatography (MDLC) coupled with electrospray-ionization (ESI)-ion trap mass spectrometry (nano-ESI-MS/MS) analyses revealed the synthesis of the pheromone PlnA in sourdough *L. plantarum* DC400 [82]. The main features of its activity and the ecological relevance are described in the Sect. 3.2.2.

References

1. Hentzer M, Wu H, Andersen JB, Riedel K, Rasmussen TB, Bagge N, Schembri MA, Song Z, Kristoffersen P, Manefield M, Costerton JW, Molin S, Eberl L, Steinberg P, Kjelleberg S, Høiby N, Givskov M (2003) Attenuation of *Pseudomonas aeruginosa* virulence by quorum sensing inhibitors. EMBO J 22:3803–3815
2. Wagner VE, Gillis RJ, Iglewski B (2004) Transcriptome analysis of quorum sensing regulation and virulence factor expression in *Pseudomonas aeruginosa*. Vaccine 22:S15–S20
3. Wu L, Estrada O, Zaborina O, Bains M, Shen L, Kohler JE, Patel N, Musch MW, Chang EB, Fu Y-X, Jacobs MA, Nishimura MI, Hancock REW, Turner JR, Alverdy JC (2005) Recognition of host immune activation by *Pseudomonas aeruginosa*. Science 309:774–777
4. Nouwens AS, Beatson SA, Whitchurch CB, Walsh BJ, Schweizer HP, Mattick JS, Cordwell SJ (2003) Proteome analysis of extracellular proteins regulated by the las and rhl quorum sensing system in *Pseudomonas aeruginosa* PAO1. Microbiology 149:1311–1322
5. Braun P, de Groot A, Bitter W, Tommassen J (1998) Secretion of elastinolytic enzymes and their propeptides by *Pseudomonas aeruginosa*. J Bacteriol 180:3467–3469
6. Folders J, Tommassen J, Van Loon LC, Bitter W (2000) Identification of a chitin-binding protein secreted by *Pseudomonas aeruginosa*. J Bacteriol 182:1257–1263
7. Nakao H, Watanabe H, Nakayama S, Takeda T (1995) yst gene expression in *Yersinia enterocolitica* is positively regulated by a chromosomal region that is highly homologous to *Escherichia coli* host factor 1 gene (hfq). Mol Microbiol 18:859–865
8. Robertson GT, Loop RM Jr (1999) The *Brucella abortus* host factor I (HF-I) protein contributes to stress resistance during stationary phase and is a major determinant of virulence in mice. Mol Microbiol 34:690–700
9. Coulthurst SJ, Monson RE, Salmond GPC (2008) Quorum sensing in the soft-rot Erwinias. In: Winans SC, Bassler BL (eds) Chemical communication among bacteria. ASM Press, Washington, DC, p 185
10. Lewenza S, Conway B, Greenberg EP, Sokol PA (1999) Quorum sensing in *Burkholderia cepacia*: identification of the LuxRI homologs CepRI. J Bacteriol 181:748–756
11. Gotschlich A, Huber B, Geisenberger O, Tögl A, Steidle A, Riedel K, Hill P, Tümmler B, Vandamme P, Middleton B, Camara M, Williams P, Hardman A, Eberl L (2000) Synthesis of multiple N-acyl-homoserine lactones is wide-spread among the members of the *Burkholderia cepacia* complex. Syst Appl Microbiol 24:1–14
12. Huber B, Riedel K, Hentzer M, Heydorn A, Gotschlich A, Givskov M, Molin S, Eberl L (2001) The cep quorum sensing system of *Burkholderia cepacia* H111 controls biofilm formation and swarming motility. Microbiology 147:2517–2528
13. Lewenza S, Sokol PA (2001) Regulation of ornibactin biosynthesis and N-acyl-L-homoserine lactone production by CepR in *Burkholderia cepacia*. J Bacteriol 183:2212–2218
14. Riedel K, Aravalo-Ferro C, Reil G, Gorg A, Lottspeich F, Eberl L (2003) Analysis of the quorum sensing *Burkholderia cepacia* H111 by proteomics. Electrophoresis 24:740–750
15. Hanna SL, Sherman NE, Kinter MT, Goldberg JB (2000) Comparison of proteins expressed by *Pseudomonas aeruginosa* strains representing initial and chronic isolates from a cystic fibrosis

patient: an analysis by 2-D gel electrophoresis and capillary column liquid chromatography-tandem mass spectrometry. Microbiology 146:2495–2508

16. Sauer K, Camper AK, Ehrlich GD, Costerton JW, Davies DG (2002) *Pseudomonas aeruginosa* displays multiple phenotypes during development as a biofilm. J Bacteriol 184:1140–1154

17. Shao CP, Hor LI (2000) Metalloprotease is not essential for *Vibrio vulnificus* virulence in mice. Inf Imm 68:3569–3573

18. Jeong HS, Rhee JE, Lee JH, Choi HK, Kim DI, Lee MH, Park S-J, Choi SH (2003) Identification of *Vibrio vulnificus* Irp and its influence on survival under various stresses. J Microbiol Biotechnol 13:159–163

19. Shin NR, Lee DY, Yoo HS (2007) Identification of quorum sensing-related regulons in *Vibrio vulnificus* by two-dimensional gel electrophoresis and differentially displayed reverse transcriptase PCR. FEMS Immunol Med Microbiol 50:94–103

20. Miyoshi N, Shinoda S (1997) Bacterial metalloprotease as the toxic factor in infection. J Toxicol Toxin Rev 16:177–194

21. Sgarrella F, Poddie FP, Meloni MA, Sciola L, Pippia P, Tozzi MG (1997) Channelling of deoxyribose moiety of exogenous DNA into carbohydrate metabolism: role of deoxyriboaldolase. Comp Biochem Physiol B Biochem Mol Biol 117:253–257

22. Di Cagno R, De Angelis M, Calasso M, Gobbetti M (2011) Proteomics of the bacterial crosstalk by quorum sensing. J Proteomics 74:19–34

23. Lereclus D, Agaisse H, Gominet M, Salamitou S, Sanchis V (1996) Identification of a Bacillus thuringiensis gene that positively regulates transcription of the phosphatidylinositol-specific phospholipase C gene at the onset of the stationary phase. J Bacteriol 178:2749–2756

24. Rasko DA, Altherr MR, Han CS, Ravel J (2005) Genomics of the *Bacillus cereus* group of organisms. FEMS Microbiol Rev 29:303–329

25. Dixon TC, Meselson M, Guillemin J, Hanna PC (1999) Anthrax. N Engl J Med 341:815–826

26. Slamti L, Lereclus D (2002) A cell-cell signaling peptide activates the PlcR virulence regulon in bacteria of the *Bacillus cereus* group. EMBO J 21:4550–4559

27. Ohtani K, Hayashi H, Shimizu T (2002) The luxS gene is involved in cell-cell signalling for toxin production in *Clostridium perfringens*. Mol Microbiol 44:171–179

28. Donlan RM (2002) Biofilms: microbial life on surfaces. Emerging Infect Dis 8:881–890

29. Flemming HC, Wingender J (2001) Relevance of microbial extracellular polymeric substances (EPSs)-part I: structural and ecological aspects. Water Sci Technol 43:1–8

30. Sutherland IW (2001) Biofilm exopolysaccharides: a strong and sticky framework. Microbiology 147:3–9

31. Korber DR, Lawrence JR, Lappin-Scott HM, Costerton JW (1995) Growth of microorganisms on surfaces. In: Lappin-Scott HM, Costerton JW (eds) Microbial biofilms, plant and microbial biotechnology research, vol 5. Cambridge University Press, Cambridge, UK, p 15

32. Ruiz LM, Valenzuela S, Castro M, Gonzalez A, Frezza M, Soulère L, Rohwerder T, Queneau Y, Doutheau A, Sand W, Jerez CA, Guiliani N (2008) AHL communication is a widespread phenomenon in biomining bacteria and seems to be involved in mineral-adhesion efficiency. Hydrometallurgy 94:133–137

33. Waters CM, Bassler BL (2005) Quorum sensing: cell-to-cell communication in bacteria. Annu Rev Cell Dev Biol 21:319–346

34. von Bodman SB, Majerczak DR, Coplin DL (1998) A negative regulator mediates quorum sensing control of exopolysaccharide production in *Pantoea stewartii* subsp. *stewartii*. Proc Natl Acad Sci USA 95:7687–7692

35. Rivas M, Seeger M, Holmes DS, Jedlicki E (2005) A Lux-like quorum sensing system in the extreme acidophile *Acidithiobacillus ferrooxidans*. Biol Res 38:283–297

36. Miller MB, Bassler BL (2001) Quorum sensing in bacteria. Annu Rev Microbiol 55:165–199

37. Cheirsilp B, Shoji H, Shimizu H, Shioya S (2003) Interactions between *Lactobacillus kefiranofaciens* and *Saccharomyces cerevisiae* in mixed culture for kefiran production. J Biosci Bioeng 96:279–284

38. Chao L, Levin BR (1981) Structured habitats and the evolution of anticompetitor toxins in bacteria. Proc Natl Acad Sci USA 78:6324–6328

39. Hooshangi S, Bentley WE (2008) From unicellular properties to multicellular behavior: bacteria quorum sensing circuitry and applications. Curr Opin Biotechnol 19:550–555

40. Micheli L, Uccelletti D, Palleschi C, Crescenzi V (1999) Isolation and characterisation of a ropy *Lactobacillus* strain producing the exopolysaccharide kefiran. Appl Environ Microbiol 53:69–74

41. Mercier KA, Cort JR, Kennedy MA, Lockert EE, Ni S, Shortridge MD, Powers R (2009) Structure and function of *Pseudomonas aeruginosa* protein PA1324 (21–170). Prot Sci 18:606–618

42. Cramton SE, Gerke C, Schnell NF, Nichols WW, Gotz F (1999) The intercellular adhesion (ica) locus is present in *Staphylococcus aureus* and is required for biofilm formation. Infect Immun 67:5427–5433

43. Dobinsky S, Kiel K, Rohde H, Bartscht K, Knobloch JKM, Horstkotte MA, Mack D (2003) Glucose-related dissociation between icaADBC transcription and biofilm expression by *Staphylococcus epidermidis*: evidence for an additional factor required for polysaccharide intercellular adhesin synthesis. J Bacteriol 185:2879–2886

44. Gerke C, Kraft A, Sussmuth R, Schweitzer O, Gotz F (1998) Characterization of the N-acetylglucosaminyltransferase activity involved in the biosynthesis of the *Staphylococcus epidermidis* polysaccharide intercellular adhesin. J Biol Chem 273:18586–18593

45. Shanks RMQ, Donegan NP, Graber ML, Buckingham SE, Zegans ME, Cheung AL, O'Toole GA (2005) Heparin stimulates *Staphylococcus aureus* biofilm formation. Infect Immun 73:4596–4606

46. Plotkowski MC, Costa AO, Morandi V, Barbosa HS, Nader HB, De Benizmann S, Puchelle E (2001) Role of heparan sulphate proteoglycans as potential receptors for non-piliated *Pseudomonas aeruginosa* adherence to non-polarised airway epithelial cells. J Med Microbiol 50:183–190

47. Read RR, Costerton JW (1987) Purification and characterization of adhesive exopolysaccharides from *Pseudomonas putida* and *Pseudomonas fluorescens*. Can J Microbiol 33:1080–1090

48. Prouty AM, Schwesinger WH, Gunn JS (2002) Biofilm formation and interaction with the surfaces of gallstones by *Salmonella* spp. Infect Immun 70:2640–2649

49. Sonck KAJ, Kint G, Schoofs G, Wauden CV, Vanderleyden J, De Keersmaecker SCJ (2009) The proteome of *Salmonella typhimurium* grown under in vivo-mimicking conditions. Proteomics 9:565–579

50. Taga ME, Semmelhack JL, Bassler BL (2001) The LuxS-dependent autoinducer AI-2 controls the expression of an ABC transporter that functions in AI-2 uptake in *Salmonella typhimurium*. Mol Microbiol 42:777–793

51. Groisman EA (2001) The pleiotropic two-component regulatory system PhoP-PhoQ. J Bacteriol 183:1835–1842

52. Agudo D, Mendoza MT, Castañares C, Nombela C, Rotger R (2004) A proteomic approach to study *Salmonella typhi* periplasmic proteins altered by a lack of the DsbA thiol: disulfide isomerise. Proteomics 4:355–363

53. Stark RM, Gerwig GJ, Pitman RS, Potts LF, Williams NA, Greenman J, Weinzweig IP, Hirst TR, Millar MR (1999) Biofilm formation by *Helicobacter pylori*. Lett Appl Microbiol 28:121–126

54. Romling U (2005) Characterization of the rdar morphotype, a multicellular behaviour in *Enterobacteriaceae*. Cell Mol Life Sci 62:1234–1246

55. Parkhill J, Wren BW, Mungall K, Ketley JM, Churcher C, Basham D, Chillingworth T, Davies RM, Feltwell T, Holroyd S, Jagels K, Karlyshev AV, Moule S, Pallen MJ, Penn CW, Quail MA, Rajandream MA, Rutherford KM, van Vliet AH, Whitehead S, Barrell BG (2000) The genome sequence of the food-borne pathogen *Campylobacter jejuni* reveals hypervariable sequences. Nature 403:665–668

56. Svensson SL, Davis LM, Mac Kichan JK, Allan BJ, Pajaniappan M, Thompson SA, Gaynor EC (2009) The CprS sensor kinase of the zootic pathogen *Campylobacter jejuni* influences biofilm formation and is required for optimal chick colonization. Mol Microbiol 71:253–272

57. Sturme MHJ, Nakayama J, Molenaar D, Murakami Y, Kunugi R, Fujii T, Vaughan EE, Kleerebezem M, de Vos WM (2005) An agr-like two-component regulatory system in *Lactobacillus plantarum* is involved in production of a novel cyclic peptide and regulation of adherence. J Bacteriol 187:5224–5235

58. Dufour P, Jarraud S, Vandenesch F, Greenland T, Novick RP, Bes M, Etienne J (2002) High genetic variability of the agr locus in *Staphylococcus species*. J Bacteriol 184:1180–1186

59. Nakayama J, Kariyama R, Kumon H (2002) Description of a 23.9-kilobase chromosomal deletion containing a region encoding fsr genes which mainly determines the gelatinase-negative phenotype of clinical isolates of *Enterococcus faecalis* in urine. Appl Environ Microbiol 68:3152–3155

60. Autret N, Raynaud C, Dubail I, Berche P, Charbit A (2003) Identification of the agr locus of *Listeria monocytogenes*: role in bacterial virulence. Infect Immun 71:4463–4471

61. De Vos WM, Bron PA, Kleerebezem M (2004) Post-genomics of lactic acid bacteria and other food-grade bacteria to discover gut functionality. Curr Opin Biotechnol 15:86–93

62. Nissen-Meyer J, Nes IF (1997) Ribosomally synthesized antimicrobial peptides: their function, structure, biogenesis, and mechanism of action. Arch Microbiol 167:67–77

63. Nes IF, Diep DB, Havarstein LS, Brurberg MB, Eijsink V, Holo H (1996) Biosynthesis of bacteriocins in lactic acid bacteria. Ant Van Leeuw 70:113–128

64. Klaenhammer TR (1993) Genetics of bacteriocins produced by lactic acid bacteria. FEMS Microbiol Rev 12:39–86

65. Cotter PD, Hill C, Ross RP (2005) Bacteriocins: developing innate immunity for food. Nat Rev Microbiol 3:777–788

66. Klarin B, Johansson ML, Molin G, Larsson A, Jeppsson B (2005) Adhesion of the probiotic bacterium *Lactobacillus plantarum* 299v onto the gut mucosa in critically ill patients: a randomised open trial. Crit Care 9:R285–R293

67. Hansen JN (1994) Nisin as a model food preservative. Crit Rev Food Sci Nutr 34:69–93

68. Kleerebezem M, Quadri LE (2001) Peptide pheromone-dependent regulation of antimicrobial peptide production in Gram-positive bacteria: a case of multicellular behaviour. Peptides 22:1579–1596

69. Kuipers OP, Beerthuizen MM, de Ruyter PGGA, Luesink EJ, de Vos WM (1995) Autoregulation of nisin biosynthesis in *Lactococcus lactis* by a signal transduction. J Biol Chem 270:27299–27304

70. Gutowski-Eckel Z, Klei C, Siegers K, Bhom K, Hammalmen M, Entian KD (1994) Growth phase-dependent regulation and membrane localization of SpaB, a protein involved in biosynthesis of the lantibiotic subtilin. Appl Environ Microbiol 60:1–11

71. Heinzmann S, Entian KD, Stein T (2006) Engineering *Bacillus subtilis* ATCC 6633 for improved production of the lantibiotic subtilin. Appl Microbiol Biotechnol 69:532–536

72. Fuchs SW, Jaskolla TW, Bochmann S, Kötter P, Wichelhaus T, Karas M, Stein T, Entian K-D (2011) Entianin, a novel subtilin-like lantibiotic from *Bacillus subtilis* subsp. *spizizenii* DSM 15029 T with high antimicrobial activity. Appl Environ Microbiol 77:1698–1707

73. Stein T, Borchert S, Conrad B, Feesche J, Hofemeister B, Hofemeister J, Entian KD (2002) Two different lantibiotic-like peptides originate from the ericin gene cluster of *Bacillus subtilis* A1/3. J Bacteriol 184:1703–1711

74. Hols P, Hancy F, Fontaine L, Grossiord B, Prozzi D, Leblond-Bourget N, Decaris B, Bolotin A, Delborme C, Ehrlich SD, Guèdon E, Monnet V, Renault P, Kleerebezem M (2005) New insights in the molecular biology and physiology of *Streptococcus thermophilus* revealed by comparative genomics. FEMS Microbiol Rev 29:435–463

75. Maldonado-Barragan A, Ruiz-Barba JL, Jimenez-Diaz R (2009) Knockout of three-component regulatory systems reveals that the apparently constitutive plantaricin production phenotype

shown by *Lactobacillus plantarum* on solid medium is regulated via quorum sensing. Int J Food Microbiol 130:35–42

76. Quadri LEN, Kleerebezem M, Kuipers OP, De Vos WM, Roy KL, Vederas JC, Stiles ME (1997) Characterization of a locus from *Carnobacterium piscicola* LV17B involved in bacteriocin production and immunity: evidence for global inducer-mediated transcriptional regulation. J Bacteriol 179:6163–6171

77. Diep DB, Håvarstein LS, Nes IF (1996) Characterization of the locus responsible for the bacteriocin production in *Lactobacillus plantarum* C11. J Bacteriol 178:4472–4483

78. Eijsink VGH, Brurberg MB, Middelhoven PH, Nes IF (1996) Induction of bacteriocin production in *Lactobacillus sake* by a secreted peptide. J Bacteriol 178:2232–2237

79. Nes IF, Eijsink VGH (1999) Regulation of group II peptide bacteriocin synthesis by quorum sensing mechanisms. In: Dunny GM, Winans SC (eds) Cell-cell signaling in bacteria. ASM Press, Washington, DC, p 175

80. Maldonado A, Ruiz-Barba JL, Jiménez-Díaz R (2004) Production of plantaricin NC8 by *Lactobacillus plantarum* NC8 is induced in the presence of different types of Gram-positive bacteria. Arch Microbiol 181:8–16

81. Diep DB, Straume D, Kjos M, Torres C, Nes IF (2009) An overview of the mosaic bacteriocin pln loci from *Lactobacillus plantarum*. Peptides 30:1562–1574

82. Di Cagno R, De Angelis M, Calasso M, Vincentini O, Vernocchi P, Ndagijimana M, De Vincenzi M, Dessi MR, Guerzoni ME, Gobbetti M (2010) Quorum sensing in sourdough *Lactobacillus plantarum* DC400: induction of plantaricin A (PlnA) under co-cultivation with other lactic acid bacteria and effect of PlnA on bacterial and Caco-2 cells. Proteomics 10:2175–2190

83. Straume D, Kjos M, Nes IF, Diep DB (2007) Quorum sensing based bacteriocin production is down-regulated by N-terminally truncated species of gene activators. Mol Genet Genomics 278:283–293

84. Hauge HH, Mantzilas D, Moll GN, Konings WN, Driessen AJ, Eijsink VG, Nissen-Meyer J (1998) Plantaricin A is an amphiphilic alpha-helical bacteriocin-like pheromone which exerts antimicrobial and pheromone activities through different mechanisms. Biochemistry 37:16026–16032

85. Nissen-Meyer J, Larsen AG, Sletten K, Daeschel M, Nes IF (1993) Purification and characterization of plantaricin A, a *Lactobacillus plantarum* bacteriocin whose activity depends on the action of two peptides. J Gen Microbiol 139:1973–1978

86. Fimland N, Rogne P, Fimland G, Nissen-Meyer J, Kristiansen PE (2008) Three-dimensional structure of the two peptides that constitute the two-peptide bacteriocin plantaricin EF. Biochim Biophys Acta 1784:1711–1719

87. Quadri LEN (2002) Regulation of antimicrobial peptide production by autoinducer-mediated quorum sensing in lactic acid bacteria. Ant Van Leeuwen 82:133–145

Chapter 3
The Behavior in Foods

3.1 Introduction

Once the language had been partly decoded and some phenotypes were directly or indirectly attributed to control by quorum sensing circuits, the dependent behavior of bacteria in foods remained to be elucidated. The language spoken between bacteria populating the same food ecosystem may condition their phenotypic traits and, consequently, their role as starter, spoilage, or pathogen microorganisms. Conversely, food matrices may contain chemical compounds that interfere with bacterial cell-to-cell communication and act as quorum quenching signals.

This chapter focuses on the most relevant evidence concerning bacterial quorum sensing mechanisms in sourdough, yogurt starter cultures, and meat and vegetable foods.

3.2 Sourdough

Sourdough is a typical example of a complex food ecosystem, where bacterial behavior and performance are influenced by interactions among coexisting species of lactic acid bacteria [1]. The use of sourdough as a natural starter for leavening of baked goods is considered to be one of the oldest biotechnological processes in food fermentations [1]. Sourdough is a mixture of flour (e.g., wheat, rye), water, and other ingredients (e.g., NaCl) that is fermented by naturally occurring lactic acid bacteria and yeasts [1, 2]. A microbial consortium, mainly consisting of obligately and/or facultatively heterofermentative lactobacilli and yeasts, usually dominates the mature sourdough [2]. The performance and stability of the mature sourdough depends on a number of factors, which include autochthonous microbiota and its metabolic activity (e.g., cofactor regeneration capability and energy synthesis from various sources), specific technology parameters (e.g., chemical and enzyme

M. Gobbetti and R. Di Cagno, *Bacterial Communication in Foods*,
SpringerBriefs in Food, Health, and Nutrition, DOI 10.1007/978-1-4614-5656-8_3,
© Marco Gobbetti and Raffaella Di Cagno 2013

composition of the flour, leavening and storage temperature, pH and redox potential, dough hydration and yield, number of sourdough refreshment steps, fermentation time between refreshments, the use of starters and/or baker's yeast), and physiological events (e.g., proto-cooperation and antagonisms) [1]. Besides, growth under microbial consortia could in some ways be considered a stress condition when compared to a mono-culture condition. Within this complex ecosystem, extracellular signaling may provide a new basis for explaining the response mechanisms to the behavior of sourdough bacteria as a consequence of heterogeneous community interactions.

3.2.1   Cross-Talk Between Sourdough Lactic Acid Bacteria

Sourdough lactic acid bacteria include bacteria that are restricted to this specific niche and have limited physiological abilities. This is the case for *Lactobacillus sanfranciscensis*, which is only found in sourdoughs [2]. Other sourdough lactic acid bacteria such as *Lactobacillus plantarum* and *Lactobacillus reuteri* are more adaptable, and are frequently isolated in various fermented foods and plant materials [3] or as natural inhabitants of the human gastrointestinal tract [4].

The proteomic approach was used as a dictionary to translate the potential cross-talk between sourdough starter lactic acid bacteria [5–7] (Fig. 3.1). Initially, the growth of the key starter bacterium *L. sanfranciscensis* CB1 under mono-culture conditions was compared to that under co-culture conditions with *L. plantarum* DC400, *Lactobacillus brevis* CR13, or *Lactobacillus rossiae* A7. This mimics some of the most frequent bacterial associations that occur during sourdough fermentation. Wheat flour hydrolyzed as the culture medium and fermentation over long time periods were chosen to resemble the chemical composition of wheat flour and the most widely used protocol of sourdough propagation [1]. The highest number of dead/damaged cells of *L. sanfranciscensis* CB1 was found in co-cultures with *L. plantarum* DC400 or *L. brevis* CR13 when the late stationary phase of growth was reached (Fig. 3.2). On the contrary, the co-cultivation with *L. rossiae* A7 did not modify the number of dead/damaged cells compared to the mono-culture. Therefore, co-cultivation with strain CR13 and, especially, DC400 might be considered as a stressing condition for *L. sanfranciscensis* CB1. Other and not easily definable factors (acidity, synthesis of antimicrobial compounds, and nutrient competition) than cell-to-cell communication might interfere with the above stress conditions. Two-dimensional electrophoresis (2-DE) analysis was carried out. The number of induced proteins markedly increased, especially when *L. sanfranciscensis* CB1 was

---

Fig. 3.1 (continued) depending on the bacterial species cross-talk. The synthesis of the pheromone plantaricin A (*PlnA*) was found only when *Lactobacillus plantarum* DC400 was co-cultured with *Lactobacillus sanfranciscensis* DPPMA174 or *Pediococcus pentosaceus* 2XA3 [7]. Reverse-phase high-pressure liquid chromatography (*RP-HPLC*), multidimensional high-performance liquid chromatography (*MDLC*) coupled with electrospray-ionization (*ESI*) ion-trap mass spectrometry (*nano-ESI-MS/MS*) analyses were used to identify PlnA

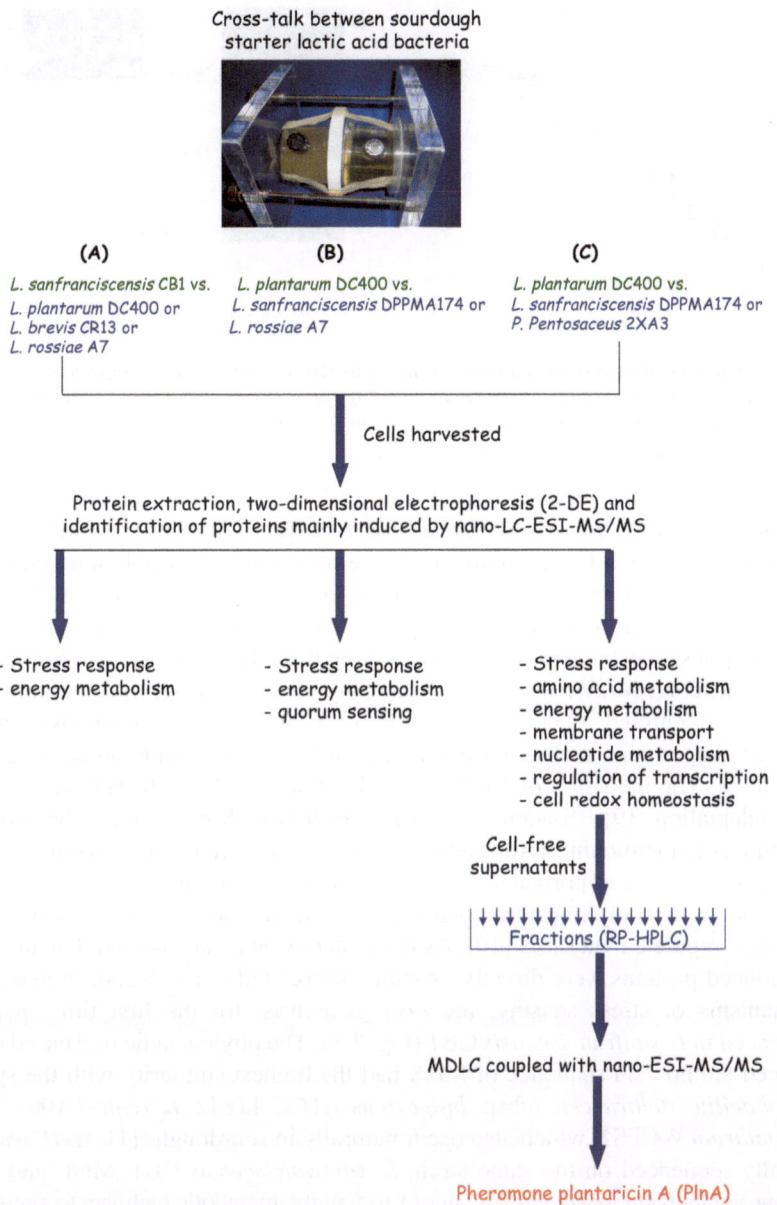

Cross-talk between sourdough
starter lactic acid bacteria

(A)

L. sanfranciscensis CB1 vs.
L. plantarum DC400 or
L. brevis CR13 or
L. rossiae A7

(B)

L. plantarum DC400 vs.
L. sanfranciscensis DPPMA174 or
L. rossiae A7

(C)

L. plantarum DC400 vs.
L. sanfranciscensis DPPMA174 or
P. Pentosaceus 2XA3

Cells harvested

Protein extraction, two-dimensional electrophoresis (2-DE) and
identification of proteins mainly induced by nano-LC-ESI-MS/MS

- Stress response
- energy metabolism

- Stress response
- energy metabolism
- quorum sensing

- Stress response
- amino acid metabolism
- energy metabolism
- membrane transport
- nucleotide metabolism
- regulation of transcription
- cell redox homeostasis

Cell-free
supernatants

Fractions (RP-HPLC)

MDLC coupled with nano-ESI-MS/MS

Pheromone plantaricin A (PlnA)

Fig. 3.1 Schematic representation of the proteomic approaches used to study the molecular mechanisms of the potential cross-talk between sourdough starter lactic acid bacteria. Co-cultures of sourdough lactic acid bacteria (the combination *A*, *B*, and *C* correspond to references [5–7] respectively) were studied during fermentation (under stirring conditions) in a double culture vessels apparatus separated by a 0.4-μm membrane filter. Cells of each strain in mono- and co-culture were harvested and two-dimensional electrophoresis (*2-DE*) profiles were compared. The proteins mainly induced were identified by Nano Liquid Chromatography Electrospray Ionization Tandem Mass Spectrometry (nano-LC-ESI-MS/MS). Induced polypeptides were related to several func tions

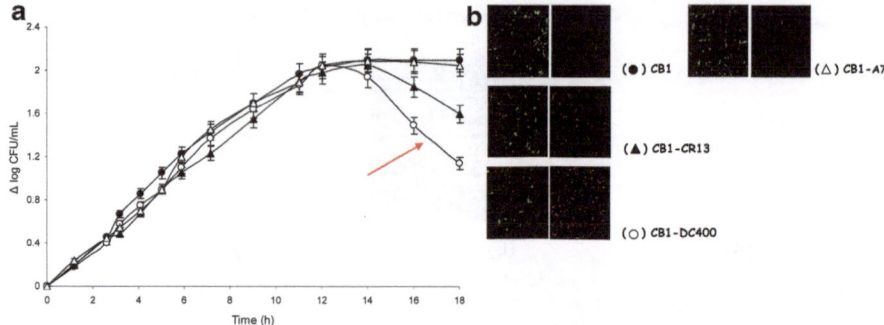

Fig. 3.2 Kinetics of growth (**a**) and fluorescing cells (**b**) of *Lactobacillus sanfranciscensis* CB1. Mono-culture (●); co-culture with *Lactobacillus plantarum* DC400 (○); co-culture with *Lactobacillus brevis* CR13 (▲); and co-culture with *Lactobacillus rossiae* A7 (△). Metabolically active and dead/damaged cells are stained green and red, respectively (Adapted from [5])

co-cultured with *L. plantarum* DC400 or *L. brevis* CR13 (47 and 45 proteins, respectively). Only a few proteins (11) were moderately induced under co-culture with *L. rossiae* A7. Twenty proteins having the highest levels of induction were identified. These had a central role in glycolysis-related machinery and, especially, in stress-response mechanisms (Fig. 3.1). Among these latter, GroES and S-adenosyl-methyltransferase MraW, which are specifically involved in cell-to-cell communication, were identified. Overall, the LuxR transcriptor of Gram-negative bacteria required chaperonins such as GroEL and GroES for folding into an active conformation [8]. The induction of GroES was also found in *L. sanfranciscensis* during acid adaptation [9]. S-adenosyl-methyltransferase MraW uses the cofactor S-adenosyl-L-methionine (AdoMet) to methylate a variety of molecular targets, thereby modulating important cellular and metabolic activities. In *Escherichia coli*, the S-adenosyl-methyltransferase (*tam* gene), which is located in the *lsr*ACDBFG operon, is regulated via LuxS [10]. As the proteome analysis showed that almost all the induced proteins were directly or indirectly related to LuxS and, in general, to mechanisms of stress sensing, the *luxS* gene was, for the first time, partially sequenced in *L. sanfranciscensis* CB1 (Fig. 3.3). The phylogenetic tree based on the deduced amino acid sequence of LuxS had the highest similarity with the species *Lactobacillus delbrueckii* subsp. *bulgaricus* ATCC 11842, *L. reuteri* 100–23 and *L. plantarum* WCFS1, which also occur naturally in sourdoughs [1]. *MetF* was also partially sequenced on the same strain *L. sanfranciscensis* CB1. MetF and MetE enzymes, which are located upstream of LuxS in the metabolic pathway to synthesize signaling molecules (see Sect. 1.5), are indispensable for generation of methionine, which is a part of S-adenosylmethionine [11]. The co-cultivation of *L. sanfranciscensis* CB1 with other sourdough lactobacilli somewhat induced the LuxS-mediated signaling. Indeed, the expression of *luxS* during the exponential phase of growth was higher when *L. sanfranciscensis* CB1 was co-cultured with *L. plantarum* DC400 and *L. brevis* CR13 compared to mono- or co-culture with *L. rossiae* A7 (Fig. 3.3). Overall, all LuxS-containing bacteria synthesize the dihydroxy-2,3-pentanedione

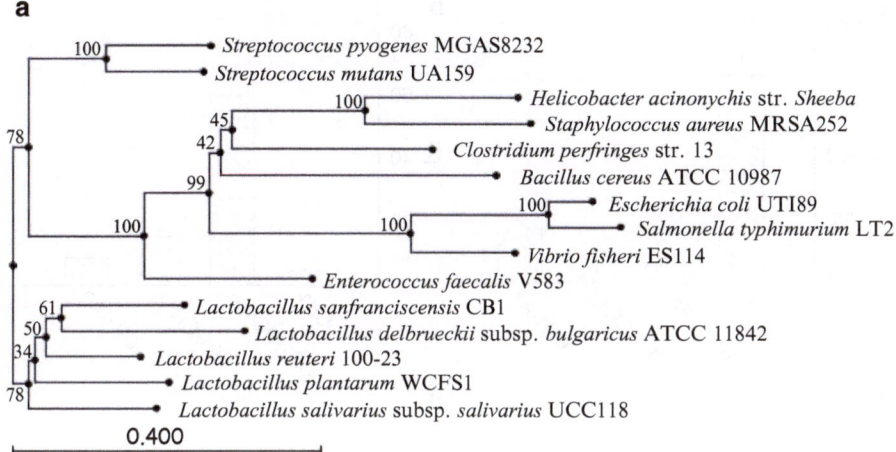

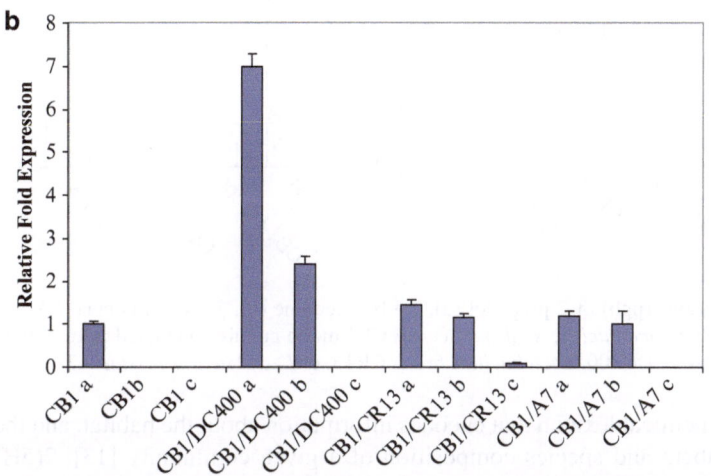

Fig. 3.3 Phylogenetic tree based upon the neighbor-joining method of deduced partial LuxS sequences. Horizontal bar represents 1 % sequence divergence. Numbers indicate bootstrap value branch points (**a**) Expression of the LuxS gene of *Lactobacillus sanfranciscensis* CB1 in monoculture (*CB1*) and co-cultures with *Lactobacillus plantarum* DC400, *Lactobacillus brevis* CR13, or *Lactobacillus rossiae* A7. (**b**) RT-PCR was performed after 7 [mid-exponential phase of growth, (a)], 12 [early stationary phase of growth, (b)], and 18 [late-stationary phase of growth, (c)] h of incubation at 30 °C (Adapted from [5])

(DPD) precursor but it is likely that DPD may cyclize to a variety of furanones (see Sect. 1.5). The evolved biological function of a number of furanone analogues seems to be acting as interspecies signal molecules under several ecosystems [12]. The chemical properties of furanones are ideal for the signaling mechanisms. They are water and/or fat soluble or volatile, depending on the substituent on the central ring [12]. The variety of signaling furanones synthesized by bacteria could be the basis

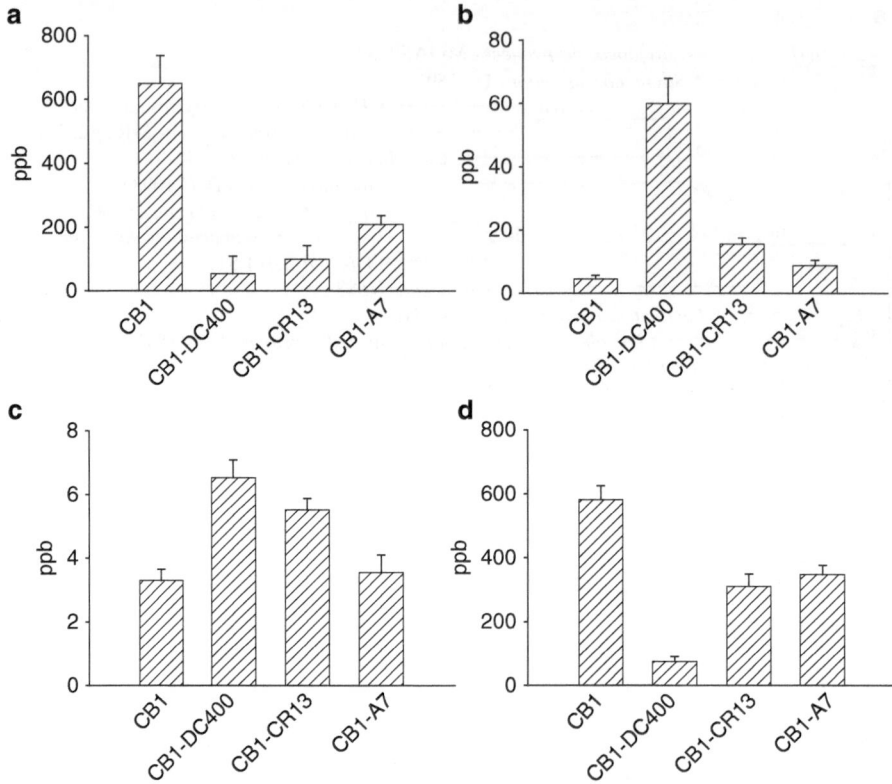

Fig. 3.4 Concentrations (ppb) of 2-propanol (**a**), 2,3-butanedoine (**b**), 3-methyl butanal (**c**), and ethyl-acetate (**d**) in the *Lactobacillus sanfranciscensis* CB1 mono-culture and mixed cultures with *Lactobacillus plantarum* DC400, *Lactobacillus brevis* CR13, or *Lactobacillus rossiae* A7

of an extensive chemical lexicon that encodes information about the habitat, and the number of members and species composition of a given community [13]. 2(5H) dihydrofuranones were hypothesized as signaling molecules of *Lactobacillus helveticus*, one of the main primary and thermophilic starters used in cheese making [14]. 2(H)dihydrofuranone-5ethyl and -5pentyl were identified as presumptive signaling molecules synthesized by *L. sanfranciscensis* CB1 when co-cultured with *L. brevis* CR13 and, especially, *L. plantarum* DC400. *Escherichia coli*, synthesized furanosyl borate diester in pure culture, which activates up-regulation to sense the environment. The same signal is synthesized at higher levels when *E. coli* was grown in mixed culture with *Vibrio harveyi*, probably, to obtain information about cell numbers of the coexisting bacterium [13]. During sourdough fermentation, the metabolism of nitrogenous compounds and the generation of volatile compounds by sourdough lactic acid bacteria influence directly or indirectly the flavor of baked goods. As interaction and communication between lactobacilli take place during sourdough fermentation, some phenotypes were conditioned by the composition of the microbial consortium. Both the synthesis of some volatile compounds (Fig. 3.4) and peptidase activities (e.g., general aminopeptidase type N, prolidase, prolinase, dipeptidase, and tripeptidase) were influenced by the microbial association, thus mirroring the different effect of other sourdough lactobacilli on stress response and cell-to-cell communication of *L. sanfranciscensis* CB1.

While the performance of *L. sanfranciscensis* CB1 was affected by the interaction with other lactobacilli, the growth and survival of *L. plantarum* DC400 was unaffected when it was co-cultivated with *L. sanfranciscensis* DPPMA174 or *L. rossiae* A7 [6]. Nevertheless, 2-DE analysis showed that the level of protein expression increased under co-culture conditions and proceeded from the mid-exponential to early stationary phases of growth. Induced polypeptides were identified as energy metabolism related, stress proteins, quorum sensing related (adenosylmethionine synthetase, MetK), and elongation factor Tu. Proteomic adaptation as the response to coexisting and, probably, stressful bacteria culminated with the synthesis of the presumptive autoinducer 2 (AI-2) that, in turn, enhanced the bioluminescence of the indicator *V. harveyi* BAA 1117. To date a number of different signaling furanones have been identified for a variety of bacteria. Probably, a unique universal bacterial signal for interspecies communication does not exist. A number of chemical compounds with the same function may be derived either as different cyclization products of DPD or as an enzymatic step downstream of LuxS.

3.2.2 The Autoinducing Peptide PlnA from Sourdough Lactobacillus plantarum

Although the universal chemical lexicon shared by both Gram-negative and -positive bacteria involves the synthesis of AI-2 through the activity of the LuxS enzyme, not only the intra- but also the interspecies communication of Gram-positive bacteria may rely on the synthesis of post-translationally modified peptides, called autoinducing peptides (AIP) (see Sect. 1.4).

The genome of *L. plantarum* WCFS1 contains relatively high numbers of peptide-based quorum sensing two-component systems, as well as other putative quorum sensing genes [15]. Some studies [16] demonstrated that competing microorganisms might activate specific component regulatory systems, which are involved in microbial antagonism. This is the case for the plantaricin system, which is regulated through the quorum sensing pathway (see Sects. 2.4, 2.4.1, and 2.4.2) [17, 18]. Overall, the secreted pheromone plantaricin A (PlnA) serves as the tool to measure the cell density of the synthesizing culture. At a certain cell density, PlnA triggers a series of phosphorylation reactions on histidine protein kinase (HPK) and cognate cytoplasmic response regulator (RR), resulting in the phosphorylated RR. This latter binds to regulated promoters of the bacteriocin regulon and activates all genes involved in the bacteriocin biosynthesis (e.g., plantaricins EF, JK, NC8 and J51) [18–20]. PlnA seems to have strain-specific antimicrobial activity [21]. Upon interaction with membrane lipids, PlnA assumes the membrane-induced α-helical structure.

This enables the nonchiral interaction with the target cell membrane, where PlnA binds to the receptor and mediates the pheromone effect [22]. The membrane interacting mode of action may explain why PlnA also displays antibacterial activity towards sensitive strains.

As shown for other strains of *L. plantarum* isolated from different food ecosystems [17, 21, 23], sourdough *L. plantarum* DC400 synthesized the autoinducer peptide plantaricin PlnA either under mono- or co-culture conditions. The biosynthesis of PlnA was variously stimulated depending on the microbial partner [7] (Fig. 3.5). Compared to mono-culture, co-cultivation of *L. plantarum* DC400 with several species of sourdough lactic acid bacteria did not cause variations of the concentration of PlnA. On the contrary, the partner *Pediococcus pentosaceus* 2XA3 and, especially, *L. sanfranciscensis* DPPMA174 induced the highest biosynthesis of PlnA, which, in turn, determined lethal conditions for it. In agreement with previous studies [5, 6], co-cultivation with strain DC400 might be considered as a stressing condition, especially for *L. sanfranciscensis* DPPMA174. Notwithstanding other mechanisms of interspecies cell-to-cell communication (e.g., LuxS mediated) and not excluding other inhibitory factors [6], PlnA determined a proteomic response in *L. sanfranciscensis* DPPMA174. The up-regulation of 31 proteins related to stress response, amino acid metabolism, energy metabolism, membrane transport, nucleotide metabolism, regulation of transcription, and cell redox homeostasis was found. At the same time, other proteins such as the cell division protein FtsZ, glutathione reductase (LRH_11212), and response regulator rrp11 were down-regulated. Although pheromone activity has the cell membrane as the main target, PlnA seemed also to interfere with the global cell metabolism of *L. sanfranciscensis* DPPMA174. Phenotypic traits such as the synthesis of volatile organic compounds, which are responsible for the sensory properties of sourdough baked goods [1], were influenced by the microbial association. Compared to mono-cultures, the stressful co-culture between *L. plantarum* DC400 and *L. sanfranciscensis* DPPMA174 was characterized by the marked decrease of the concentration of diacetyl, acetoin, ethylacetate, and furanone A. On the contrary, the signaling molecule furanon B, and heptadecane and decanoic acids increased or were only synthesized under co-culture conditions. The robustness of *L. plantarum* is already known within several food ecosystems. Notwithstanding other regulatory factors such as acidity, nutrient competition, synthesis of diacetyl- and LuxS-mediated compounds, pheromone PlnA could play a central role in the regulation of the competitive advantage of this bacterium within various food ecosystems.

Although the phenomenon of cell-to-cell communication between prokaryotic and eukaryotic cells is already known [24], limited attention has been paid to interactions between quorum sensing molecules (e.g., peptide pheromones) and human intestinal mucosa. It seems that PlnA has the capacity to prevent human intestinal cell damage and to enhance barrier functions (see Sect. 4.3).

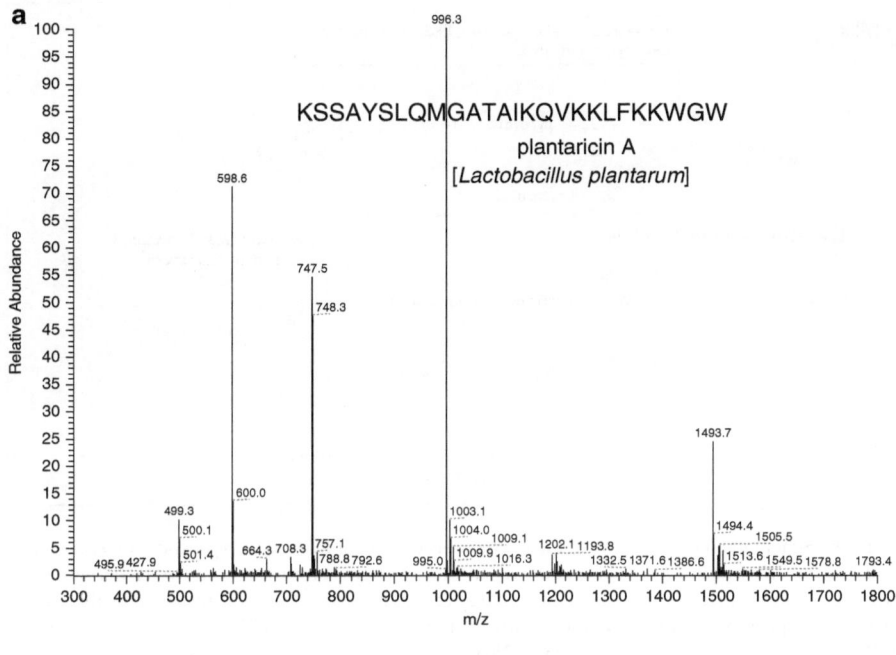

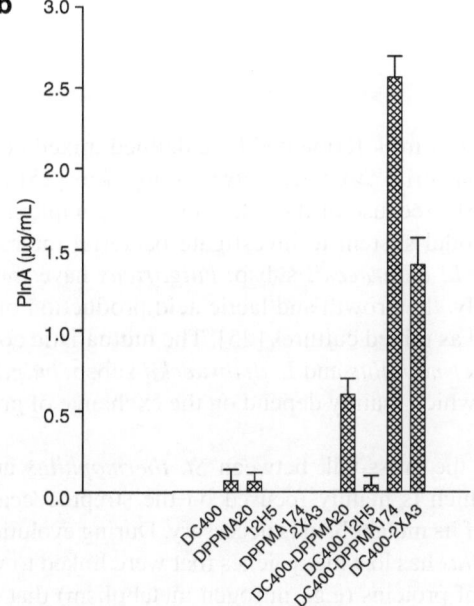

Fig. 3.5 Electrospray-ionization (ESI) ion-trap MS chromatograms of acquisition with m/z ratios related to plantaricin A (PlnA) (**a**) and concentration (μg/mL) of PlnA synthesized by mono-culture of *Lactobacillus plantarum* DC400, *L. plantarum* DPPMA20, *Lactobacillus pentosus* 12H5, *Lactobacillus sanfranciscensis* DPPMA174, *Pediococcus pentosaceus* 2XA3 and co-culture of *L. plantarum* DC400 with *L. plantarum* DPPMA20, *L. pentosus* 12H5, *L. sanfranciscensis* DPPMA174 or *P. pentosaceus* 2XA3 (**b**) (Adapted from [7])

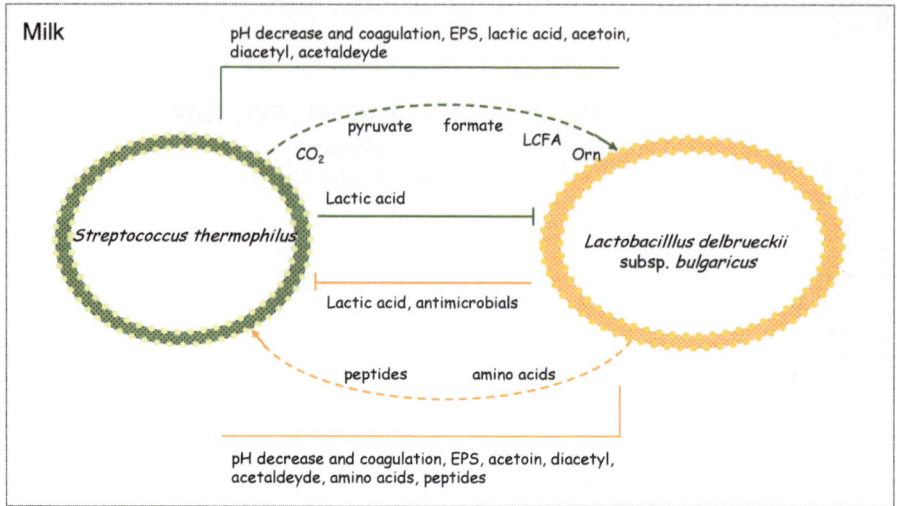

Fig. 3.6 Schematic representation of the interactions that occur between *Streptococcus thermophilus* and *Lactobacillus delbrueckii* subsp. *bulgaricus* during yoghurt fermentation. *Dotted arrows*, positive interactions; *interrupted arrows*, negative interactions; and *solid lines*, interactions that do not specifically promote or decrease the growth of the other species. LCFA, long-chain fatty acids; EPS, exopolysaccharides (Adapted from [25])

3.3   Yoghurt

Yoghurt is the product of milk fermented by a defined mixed culture of two thermophilic lactic acid bacteria: *Streptococcus thermophilus* [25] and *L. delbrueckii* subsp. *bulgaricus* [25]. Because of the relatively low complexity, yoghurt represents an attractive model system to investigate bacterial interactions. Although *St. thermophilus* and *L. delbrueckii* subsp. *bulgaricus* have the capacity to ferment milk individually, the growth and lactic acid production of both species are stimulated when used as mixed cultures [25]. The mutualistic coexistence (proto-cooperation) of *St. thermophilus* and *L. delbrueckii* subsp. *bulgaricus* is based on various interactions, which mainly depend on the exchange of growth-stimulating factors (Fig. 3.6).

Before describing the cross-talk between *St. thermophilus* and *L. delbrueckii* subsp. *bulgaricus*, which is mainly focused on the streptococcal species, a brief introductory outline of its metabolism is necessary. During evolution and adaptation to milk, *St. thermophilus* has lost many genes that were linked to virulence, and has kept an essential set of proteins (e.g., nitrogen metabolism) that are indispensable for growth in milk [26]. Consequently, *St. thermophilus* is adapted to grow on lactose, rapidly converting it into lactate. Lactose is transported into the cell through lactose permease (LacS), which operates as a galactoside proton symport system or as a lactose-galactose antiporter [27]. After uptake, lactose is hydrolyzed by an intracellular β-galactosidase. Most of the *St. thermophilus* strains only metabolized the glucose moiety of lactose, with the galactose moiety excreted into the environment [28]. Milk is poor in free amino acids (FAA) and short peptides [29], therefore, the optimal growth of *St. thermophilus* relies on the hydrolysis of caseins, and internalization and degradation of the resulting peptides or on de novo synthesis of FAA [30]. After inoculation in milk, *St. thermophilus* grows exponentially using the few available FAA and oligopeptides. When FAA become limiting, the culture enters into a nonexponential phase of growth, in which the synthesis of extracellular proteases occurs. The proteolytic system supplies sufficient peptides for the second exponential phase of growth, which shows a lower rate than the previous one [31]. The hydrolysis of milk caseins of many lactic acid bacteria mostly depends on the activity of the cell wall-associated proteinase. Few strains of *St. thermophilus* possess this proteinase [32]. The second exponential phase of growth of *St. thermophilus* clearly differs between mono- and co-cultures with *L. delbrueckii* ssp. *bulgaricus*. During mono-culture, *St. thermophilus* encounters conditions that hamper growth. Under co-culture conditions, it overcomes this effect.

3.3.1 Cross-Talk Between *Streptococcus thermophilus* and *Lactobacillus delbrueckii* subsp. *bulgaricus*

Proteomic and transcriptomic approaches were combined to investigate the potential involvement of cross-talk between the two thermophilic lactic acid bacteria during milk fermentation [33]. The comparison of the proteome of *St. thermophilus* revealed that 27 proteins were down- or up-regulated under mono- and co-culture with *L. delbrueckii* ssp. *bulgaricus*. These proteins concerned amino acid biosynthesis, carbon and purine-pyrimidine metabolisms, response regulator RR05, and other unknown functions. In particular, proteins related to the synthesis of cysteine from glyceraldehydes 3-P (SerA, Cyse2 and CysM1), trans-sulfuration and sulfhydrylation pathways (MetA, MetB1, Stu0353 and CysD), and conversion of homocysteine to methionine (MetE and MetF), and methionine to cysteine (MetK, CysM2 and MetB2), which are involved in the activated methyl cycle (AMC), were up-regulated. As amino acid and peptide transporters, and the sulfur amino acid metabolism were induced both under mono- and co-culture, one of the main literature hypotheses, which concerned the fulfillment of peptides/amino acid requirements of *St. thermophilus* by *L. delbrueckii* ssp. *bulgaricus*, was contradicted. The switch-on of the sulfur amino acid biosynthesis pathways of *St. thermophilus* during co-culture suggested that the stimulatory effect of *L. delbrueckii* ssp. *bulgaricus* is likely to result from other and more complex exchange between the two species. Conversely, the level of expression of RR05 response regulator, which is a part of the two-component regulatory system (2CRS), decreased during the second exponential phase of growth of *St. thermophilus*. This indicated that a regulatory event took place and that the activity of 2CRS might be required by *St. thermophilus* for rapid/normal growth. All available *St. thermophilus* genome sequences (strains LMG18311, CNRZ1066, and LMD-9) encode 6–8 complete and potentially functional 2CRS, including 14–16 proteins, with typical HPK and RR organization [30]. A strict correlation between number of 2CRS genes and genome size is usually found [34]. Completely sequenced prokaryotic genomes indicate that microorganisms living in stable environments use relatively simple signal-transduction systems, whereas microorganisms surviving in diverse ecological niches typically encode complex sensory systems [35]. Compared to other *Streptococcaceae*, which have over 40 2CRS proteins, *St. thermophilus* LMD-9, which has a small genome (ca. 1850 Mb), possesses only 14 2CRS proteins (8 RR and 6 HPK). Compared to strains LMG18311 and CNRZ1066, *St. thermophilus* LMD-9 had lost two HPK. All RR genes of strain LMD-9 were expressed during growth in milk but at various levels [36]. This suggested that cultivation in milk is not optimal for the high levels of expression of all RR genes. During all phases of growth, the most highly expressed genes were *rr01*, *rr05*, and *rr08*, which correspond to the most highly conserved RR in Firmicutes. Apart from the environmental conditions (e.g., chemically defined medium, milk, mixed culture with *L. delbrueckii* subsp. *bulgaricus* or rat digestive tract), 2CRS based on RR01 (2CRS01) and RR05 (2CRS05) were systematically detected in *St. thermophilus* [33, 37, 38]. These findings suggested that the above-mentioned 2CRS are essential for bacterial viability, and perform basal functions under all environmental conditions, including those characterizing the yoghurt. When in the presence of its biotic constraint *L. delbrueckii* ssp. *bulgaricus*, *St. thermophilus* modulates the expression of *rr01*, *rr02*, *rr05*, and *rr09* genes. The regulation is strain-dependent. 2-CRS based on RR09 (2CRS09) controls the synthesis of the bacteriocin thermophilin 9 of *St. thermophilus* LMD-9. This may correspond to a nonspecific defensive phenomenon due to the coexistence of another bacterium in the close environment. The 2CRS02 regulates the synthesis of the virulence and toxin factors. All the above findings together may suggest that *St. thermophilus* senses the presence of its yogurt partner and, consequently, sets up regulatory responses.

3.3.2 The Presumptive New Language Spoken by Streptococcus thermophilus

Besides the 2CRS process (see Sect. 1.4), which involves the detection of AIP outside the cell via HPK, cell-to-cell communication of Gram-positive bacteria may rely on sensing signaling molecules inside the cell. This occurs after internalization by an oligopeptide permease transport system (Opp or Ami), which is a member of the ubiquitous ATP-binding cassette super-family (ABC transporters) [39]. Once internalized, the pheromone interacts with transcriptional regulators (e.g., Rgg) or Rap protein (RNPP family), thereby modifying their activity and, consequently, the expression of the target genes [40]. Transcriptional regulators Rgg are involved in several physiological functions such as the expression of glucosyltransferases, stress response, synthesis of bacteriocins, and virulence. Overall, the regulatory mechanism by Rgg proteins is complex and has still to be completely elucidated.

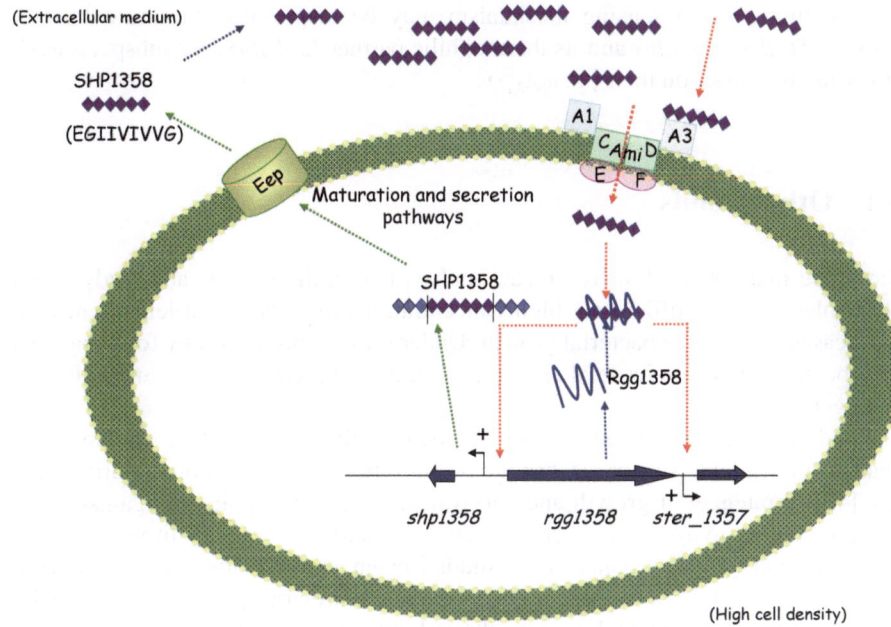

Fig. 3.7 Schematic representation of the SHP/Rgg 1358 quorum sensing mechanism of *Streptococcus thermophilus* LMD-9. The signal is encoded by the *shp1358* gene. The pheromone, a small hydrophobic peptide called SHP1358 (EGIIVIVVG) is produced by C-terminal cleavage of a 23-amino-acid peptide precursor. At high cell density, the secreted signal is sensed by the lipoprotein AmiA3 and internalized into the cell by the AmiCDEF transporter. Then the signal interacts with the regulatory protein Rgg1358 leading to the activation of their transcription. At low cell density, the pheromone SHP1358 is present at an insignificant level, and the protein Rgg1358 is bound alone to the promoter region of *shp1358* and *ster_1357* in an inactivated state. Eep, endopeptidase (Adapted from [43])

A novel quorum sensing mechanism, which is based on the Rgg transcriptional regulator and encoded by the *rgg1358* gene, was highlighted in *St. thermophilus* [41]. The mechanism includes the following steps: (1) secretion of the putative small hydrophobic peptide (SHP) pheromone called SHP1358 (sequence EGIIVIVVG), which is produced by C-terminal cleavage of a 23-amino-acid peptide precursor; (2) maturation of the signal SHP1358; (3) detection of the signal via Rgg at the cell density threshold; and (4) re-importation of SHP1358 into the cell through the AmiCDEF oligopeptide transporter. Then, the signal interacts with the regulatory protein Rgg1358 leading to the activation of their transcription (Fig. 3.7). It was hypothesized that the triggering of the mechanism is biomass dependent and relies on the presence of the mature form of the *shp1358* gene product (SHP1358) in the extracellular medium. At low cell density, the pheromone SHP1358 is present at insignificant levels, and the protein Rgg1358 is bound alone to the promoter region of the two target genes *shp1358* and *ster1357* under an inactive state.

Also this quorum sensing mechanism may be responsible for the cross-talk between *St. thermophilus* and its thermophilic partner *L. delbruecki* subsp. *bulgaricus*, with an impact on their phenotypes.

3.4   Other Foods

Raw food matrices such as red meat, poultry, fish, milk, sprouts and ready-to-eat vegetables contain sufficiently high concentrations of various nutrients, which in most cases allow early bacterial growth. Under these conditions, raw food matrices may be subjected to microbial spoilage, and, in some cases, to contamination by pathogens.

Food spoilage is considered an ecological phenomenon that encompasses changes in available substrates through proliferation of the microbiota during storage [42]. Strategies of growth and survival of ephemeral spoilage organisms fall into three main environmental determinants: (1) incidence of competitors; (2) stress response; and (3) disturbance (e.g., sudden event that provides newly available resources for exploitation). As already stated in the previous paragraphs, microbial adaptation to environmental stresses throughout space and during time is strictly related to microbial communication [43]. Signaling compounds such as N-acyl homoserine lactones (AHL) (AI-1) and furanosyl borate diester (AI-2) are synthesized and/or increase their concentration as the number of spoilage bacteria increases in various food ecosystems [44, 45]. These signal compounds also affect bacterial poisoning by stimulating the expression of several phenotypes such as biofilm formation and secretion of virulence factors. Some examples of food spoilage and poisoning, which are mediated by quorum sensing mechanisms, are described below.

3.4.1   Meat Spoilage and Poisoning

Microbial growth and activity are well supported by fresh meat, and chill-stored meat products may undergo deterioration due to microbial activity. Factors such as temperature, gaseous atmosphere, and pH affect the shelf life of the fresh meat. The main bacteria implicated in the spoilage of refrigerated meat are *Brochothrix thermosphacta*, *Lactobacillus* spp., *Leuconostoc* spp., *Carnobacterium* spp., *Pseudomonas* spp., and *Enterobacteriaceae* [46, 47]. The microbiota of aerobically packed meat stored at chill temperatures is dominated by *Pseudomonas* spp. In particular, *Pseudomonas fluorescens*, *Pseudomonas putida*, and *Pseudomonas fragi* may reach values of cell density of ca. 9.0 Log cfu/g at the point of spoilage. The quorum sensing system is used by *Pseudomonas* spp. to produce slime at the surface of aerobically packed meat [48]. The microbiota of vacuum packed fresh meat stored at chill temperatures typically consists of lactic acid bacteria and

Enterobacteriaceae. When spoilage occurs, *Hafnia alvei* and *Serratia* spp. dominate the *Enterobacteriaceae* microbiota. These microorganisms have the capacity to synthesize AHL to regulate the gene expression. Evidence was reported concerning the direct correlation between the presence of signaling compounds (AHL and AI-2) and the contamination by specific or ephemeral spoilage microorganisms [44, 49]. Bacterial phenotypes, which are regulated via AHL, may influence the sensory, nutritional and hygiene features of meat foods. Overall, the enzyme degradation of a few other raw matrices (e.g., milk) is accelerated through AHL regulation of hydrolytic enzymes [48]. Detectable levels of AHL were related to a certain cell density of *Enterobacteriaceae* (e.g., ca. 6.0 Log cfu/ml). This cell density, which is frequently found in spoiled chilled packed meats, is sufficient to activate quorum sensing systems, and these signaling compounds are frequently observed also in other foods (e.g., smoked fish) where *Enterobacteriaceae* grow well. Signals from one *Enterobacteriaceae* species may also induce relevant phenotypes for food spoilage in other species, which share the same environment (e.g., *Serratia proteamaculans*). The efficient growth of *P. fragi* in fresh meat does not seem to be regulated by an AHL-mediated quorum sensing system [50]. Nevertheless, *P. fragi*-induced bioluminescence in *Vibrio harveyi* BB170 through AI-2 activity and the potential role of AI-2 in the dynamic of the spoilage should not be excluded.

Lactic acid bacteria are considered specific or ephemeral spoilage microorganisms, which may contribute to the spoilage of modified atmosphere packaged meat products. AI-2 signals were proposed as potential compounds, which may be directly or indirectly involved in spoilage. The synthesis of the AI-2 signaling molecule was found in lactic acid bacteria isolated from milk, dairy products, and human or animal gastrointestinal tracts. *Leuconostoc* spp., *Leuconostoc mesenteroides*, *Leuconostoc citreum*, *Weissella viridescens*, and *Lactobacillus sakei*, which were isolated from minced beef stored under modified atmospheres at various temperatures, were screened for their capacity to exhibit AI-2-like activity. Signaling activity was found for *Leuconostoc* spp. isolates. Most of these isolates dominated during storage at 10 and 15 °C, and during initial and middle stages of storage at chill temperatures (0 and 5 °C). The synthesis of AI-2-like molecules was dependent on environmental factors such as time and temperature of storage, and growth medium. Fatty acids (e.g., linoleic, oleic, palmitic, and stearic acids) from ground beef and poultry meat as well as food additives (e.g., sodium-propionate, -benzoate, -acetate and -nitrate) may inhibit the AI-2 activity [51, 52]. AI-2-like activity was also searched for on cell-free meat extracts (CFME), from which the above lactic acid bacteria were isolated. Contrary to the results found for lactic acid bacteria, CFME exhibited low AI-2-like activity. This finding was also confirmed in beefsteak, beef patties, chicken breast and turkey patties, which showed a rather high level (from 6.0 to 8.0 Log cfu/ml) of autochthonous population [52]. As to the low AI-2 activity found in CFME, the effect of these signaling molecules on lactic acid bacteria under this ecological niche is still under debate. Nevertheless, clear evidence was found concerning the interference of quorum sensing molecules in the growth and physiological behavior of Gram-negative bacteria, which contaminate and spoil meat. Overall, it seems that the synthesis of AI-2-like molecules affects the dominance of various bacterial strains during meat storage, especially interfering with their persistence.

Campylobacter jejuni is largely found as a commensal bacterium in the cecum of chickens. During slaughter, it may have ample opportunity to disseminate onto processed meat and skin [53]. As a consequence, *C. jejuni* may contaminate poultry meat and cause gastrointestinal infections when, especially, undercooked products are consumed. Contrarily to other food-borne pathogens, *C. jejuni* does not have the capacity to multiply in poultry meat stored at refrigeration temperatures [54]. Nevertheless, it has the capacity to survive on raw chicken meat and skin at 4 and −20°C for more than 10 days [55, 56]. Quorum sensing mediation via the LuxS protein should play a key role in adaptation at low temperatures, which characterizes the processing chain of poultry meat [57]. *Escherichia coli* O157:H7 is another important causative agent of severe gastrointestinal disease in humans. This pathogen is characterized by a low infection dose, as low as 10–100 cells [58]. A large number of outbreaks by *E. coli* O157:H7 were associated with the consumption of contaminated ground beef and raw milk. *Escherichia coli* O157:H7 has the capacity to adhere, colonize, and form a biofilm on a variety of surfaces (see Sect. 2.3) [59]. The formation of biofilm on meat and poultry by *E. coli* O157:H7 may be attributed to AI-2 signals, which are responsible for the regulation of chemotaxis, flagellar synthesis and motility genes. The synthesis of biofilm and AI-2 signals by *E. coli* O157:H7 makes it even more difficult to efficiently control the cross-contamination of this strain during food processing.

Blocking/quenching of quorum sensing via degradation of signal molecules was proposed as a promising alternative to diminish bacterial virulence under various food ecosystems [60, 61] (see Sect. 5.2). Among the various enzymes that potentially may have a role in quorum quenching, lactonase, which catalyzes hydrolytic ring opening of the lactone to form an N-acyl-homoserine product, was the most studied. The lactonase gene is widely spread within food-borne strains of *Bacillus* spp. The quorum sensing signal molecule AHL synthesized by *Yersinia enterocolitica* was quenched by *Bacillus cereus* during co-cultivation under food simulating conditions (e.g., pork extract) [62]. The AHL-degrading capacity of 20 *Bacillus* sp. strains was preliminarily evaluated with both synthetic AHL molecules (N-hexanoyl-L-homoserine lactone and 3-oxo-hexanoyl-L-homoserine lactone) and AHL synthesized by *Y. enterocolitica*. About 75 % of strains showed this degrading capacity both at 30 °C and 7 °C. The degrading factor was not excreted outside the cell. AHL molecules have to diffuse inside the *Bacillus* sp. cells where inactivation takes place. Using *Y. enterocolitica* as the AHL producer and *B. cereus* as the AHL degrader model, the performance of these degrading mechanisms was confirmed under food simulating conditions (e.g., pork extract). Overall, the AHL-degrading capacities of *Bacillus* spp. may be used as a competitive advantage over bacterial competitors, and it may help this genus to dominate some ecological niches.

Other enzymes that potentially may have a role in quorum quenching are the AHL acylase, which catalyze the hydrolytic cleavage of the amide bond to form homoserine lactone and free fatty acid [63]. The most fully characterized AHL acylase is the PvdQ from *P. aeruginosa* [64]. PvdQ is synthesized as an inactive protein,

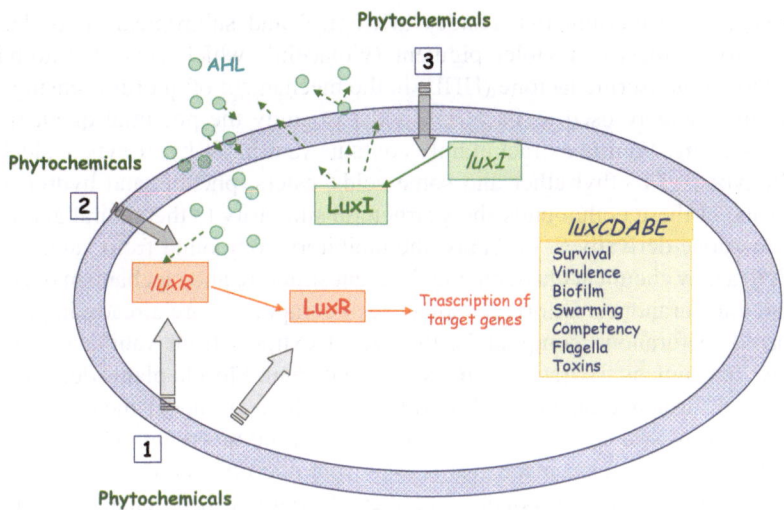

Fig. 3.8 Schematic representation of quorum sensing inhibition and/or modulation by dietary phytochemicals: (*1*) by competing directly with the LuxR family of N-acyl homoserine lactone (*AHL*) receptors; (2) by competing with the AHL molecules and/or preventing the binding of the AHL molecules to these receptors; and (*3*) by decreasing the expression of the LuxI family of synthases, thus modulating the ability of the bacteria to synthesize AHL molecules. As a consequence the phenotypes that are AHL related are inhibited or modulated

containing a signal peptide that directs export into the periplasmic space. Export is followed by intermolecular cleavage to remove the signal peptide. AHL acylase activity may be used by *P. aeruginosa* as a mechanism to detoxify the AHL compounds and to provide nutrients for growth.

3.4.2   Raw Vegetables as Sources of Quorum Sensing Inhibitors

Beyond nutrition, vegetables undoubtedly share several health-promoting features, and bacterial infections due to their consumption are more limited compared to several animal foods. As a result of coexistence over millions of years, plants may have evolved natural quorum sensing inhibitors, which should attenuate bacterial fouling and infections. Inhibition/quenching of quorum sensing activity takes place through a combination of two different mechanisms: (1) interference of phytochemicals with the activity of signal molecules; and (2) modulation of the bacterial synthesis of signal molecules via phytochemicals. An example of the model for quorum sensing inhibition through dietary phytochemicals is represented in Fig. 3.8.

A large number of plants were studied for their capacity to antagonize bacterial quorum sensing. Both cured vanilla beans (*Vanilla planifolia* Andrews) [65] and *Tremella fuciformis* [66] extracts inhibited the synthesis of violacein by the reporter strain of *Chromobacterium violaceum*. *Chromobacterium violaceum*, a soil-borne

Gram-negative bacterium that resides in tropical and subtropical areas, has the capacity to synthesize a violet pigment (violacein), which acts as autoinducer N-hexanoyl homoserine lactone (HHL) in the mechanism of quorum sensing. This bacterium is widely used as bacterial model to study the potential quenching of quorum sensing. Vanilla extract mainly contains vanillin, 4-hydroxybenzaldehyde, 4-hydroxybenzyl methyl ether and some acids, esters, phenols and hydrocarbons [65]. None of these compounds show structural similarity to the natural autoinducers or furanone derivatives. Probably, the inhibitory compound from vanilla corresponds to a new chemical class, having different structure and mechanism of activity compared to furanones. Efforts to isolate such compounds are already in progress. Compared to furanone compounds, the use of extracts from vanilla is certainly healthier for humans. Extracts from six different south Florida plants decreased the synthesis of virulence factors by *P. aeruginosa*, which are under the control of the quorum sensing system (see Sect. 2.2) The red marine macroalga *Delisea pulchra* developed natural defense mechanisms to prevent microbial colonization of its surface [67]. Overall, algae produce a number of nontoxic halogenated metabolites (e.g., brominated furanones), which effectively inhibit the synthesis of AI-2 and prevent the formation of biofilm. Broccoli (*Brassica oleracea*), as an example of soluble fiber, have numerous beneficial effects on human health. These vegetables also contain a high level of vitamin C and multiple nutrients with potent anticancer properties [68]. Broccoli extract (BE) suppressed the synthesis of AI-2 and the AI-2-mediated bacterial motility of *E. coli* O157:H7 in a dose-dependent manner. *Escherichia coli* O157:H7 harbors quorum sensing-regulated virulence genes on a pathogenicity island, termed the locus of enterocyte effacement (LEE). This promotes human intestinal colonization through the formation of lesions. The expression of the *ler* gene, which encodes the master regulator of LEE genes, was diminished in response to treatment with BE. As LEE genes are regulated through the AI-3/norepinephrine quorum sensing system [69], this suggests that BE may also target the AI-3 mechanism (see Sect. 1.6). Flavonoid compounds such as quercetin, kaempferol, and myricetin were also identified as quorum sensing inhibitors. The crude extract of *Armoracia rusticana* (horseradish) is also highly active as a quencher of the quorum sensing system of *P. aeruginosa*. 1-isothiocyanato-3-(methylsulfinyl) propane, commonly known as iberin, which is synthesized by many members of the *Brassicaceae* family, was found as a very active quorum sensing inhibitor [70]. Iberin specifically and effectively targets two of the major quorum sensing networks of *P. aeruginosa*, the LasIR and RhlIR systems. Honey is another example of a dietary source that causes the inhibition of the quorum sensing of *P. aeruginosa* [71]. Honey has been known for its medicinal uses since ancient times. Its antimicrobial properties are still not completely understood. The two largest constituents of honey are carbohydrates (ca. 81%) and water (ca. 17%). The remaining 1–2% is represented by various enzymes and miscellaneous compounds, whose chemical composition is fundamental for the bactericidal activity and varies depending on nectar source. Recently, it was shown that nonbactericidal concentrations of honey (6% or less) quenched the expression of genes for virulence factors of *P. aeruginosa*. Quorum quenching was associated with the content of sugars (mainly glucose and fructose). It was also shown that the ability of honey to counteract infections, seemed to be the result of two independent mechanisms, which act in tandem: (1) a bactericidal effect from unique molecules that is dependent on the nectar source; and (2) quorum quenching that is independent of the nectar source.

References

1. Gobbetti M, De Angelis M, Corsetti A, Di Cagno R (2005) Biochemistry and physiology of sourdough lactic acid bacteria. Trends Food Sci Technol 16:57–69
2. Gobbetti M (1998) Interactions between lactic acid bacteria and yeasts in sourdoughs. Trends Food Sci Technol 9:267–274
3. Caplice E, Fitzgerald GF (1999) Food fermentations: role of microorganisms in food production and preservation. Int J Food Microbiol 50:131–149
4. Ahrne S, Nobaek S, Jeppsson B, Adlerberth I, Wold AE, Molin G (1998) The normal *Lactobacillus* flora of healthy human rectal and oral mucosa. J Appl Microbiol 85:88–94
5. Di Cagno R, De Angelis M, Limitone A, Minervini F, Simonetti MC, Buchin S, Gobbetti M (2007) Cell-cell communication in sourdough lactic acid bacteria: a proteomic study in *Lactobacillus sanfranciscensis* CB1. Proteomics 7:2430–2446
6. Di Cagno R, De Angelis M, Coda R, Minervini F, Gobbetti M (2009) Molecular adaptation of sourdough *Lactobacillus plantarum* DC400 under co-cultivation with other lactobacilli. Res Microbiol 160:358–366
7. Di Cagno R, De Angelis M, Calasso M, Vincentini O, Vernocchi P, Ndagijimana M, De Vincenzi M, Dessi MR, Guerzoni ME, Gobbetti M (2010) Quorum sensing in sourdough *Lactobacillus plantarum* DC400: induction of plantaricin A (PlnA) under co-cultivation with other lactic acid bacteria and effect of PlnA on bacterial and Caco-2 cells. Proteomics 10:2175–2190
8. Adar YY, Simaan M, Ulitzur S (1992) Formation of the LuxR protein in the Vibrio fischeri lux system is controlled by HtpR through the GroESL proteins. J Bacteriol 174:7138–7143
9. De Angelis M, Bini L, Pallini V, Cocconcelli PS (2001) The acid-stress response in *Lactobacillus sanfranciscensis* CB1. Microbiology 147:1863–1873
10. Wang L, Li J, March JC, Valdes JJ, Bentley WE (2005) LuxS-dependent gene regulation in *Escherichia coli* K-12 revealed by genomic expression profiling. J Bacteriol 187:8350–8360
11. Schauder S, Penna L, Ritton A, Manin C, Parker F, Renauld-Mongenie G (2005) Proteomics analysis by two-dimensional differential gel electrophoresis reveals the lack of a broad response of *Neisseria meningitidis* to in vitro-produced AI-2. J Bacteriol 187:392–395
12. Slaughter JC (1999) The naturally occurring furanones: formation and function from pheromone to food. Biol Rev Camb Philos Soc 74:259–276
13. Xavier KB, Bassler BL (2005) Interference with AI-2 mediated bacterial cell–cell communication. Nature 437:750–753
14. Ndagijimana M, Vallicelli M, Cocconcelli PS, Cappa F, Patrignani F, Lanciotti R, Guerzoni ME (2006) Two 2[5 H]-furanones as possible signalling molecules in *Lactobacillus helveticus*. Appl Environ Microbiol 72:6053–6061
15. Sturme MHJ, Francke C, Siezen RJ, de Vos WM, Kleerebezem M (2007) Making sense of quorum sensing in lactobacilli: a special focus on *Lactobacillus plantarum* WCFS1. Microbiology 153:3939–3947
16. Maldonado A, Ruiz-Barba JL, Jiménez-Díaz R (2004) Production of plantaricin NC8 by *Lactobacillus plantarum* NC8 is induced in the presence of different types of Gram-positive bacteria. Arch Microbiol 181:8–16

17. Diep DB, Straume D, Kjos M, Torres C, Nes IF (2009) An overview of the mosaic bacteriocin *pln* loci from *Lactobacillus plantarum*. Peptides 30:1562–1574

18. Nes IF, Diep DB, Havarstein LS, Brurberg MB, Eijsink V, Holo H (1996) Biosynthesis of bacteriocins in lactic acid bacteria. Ant Van Leeuw 70:113–128

19. Diep DB, Havarstein LS, Nissen-Meyer J, Nes IF (1994) The gene encoding plantaricin A, a bacteriocin from *Lactobacillus plantarum* C11, is located on the same transcription unit as an agr-like regulatory system. Appl Environ Microbiol 60:160–166

20. Hauge HH, Mantzilas D, Moll GN, Konings WN, Driessen AJ, Eijsink VG, Nissen-Meyer J (1998) Plantaricin A is an amphiphilic alpha-helical bacteriocin-like pheromone which exerts antimicrobial and pheromone activities through different mechanisms. Biochemistry 37:16026–16032

21. Anderssen EL, Diep DB, Nes IF, Eijsink VG, Nissen-Meyer J (1998) Antagonistic activity of *Lactobacillus plantarum* C11: two new two-peptide bacteriocins, plantaricins EF and JK, and the induction factor plantaricin A. Appl Environ Microbiol 64:2269–2272

22. Kristiansen PE, Fimland G, Mantzilas D, Nissen-Meyer J (2005) Structure and mode of action of the membrane-permeabilizing antimicrobial peptide pheromone plantaricin A. J Biol Chem 280:22945–22950

23. Navarro L, Rojo-Bezares B, Saenz Y, Diez L, Zarazaga M, Ruiz-Larrea F, Torres C (2008) Comparative study of the pln locus of the quorum sensing regulated bacteriocin- producing *L. plantarum* J51 strain. Int J Food Microbiol 128:390–394

24. Fujiya M, Musch MW, Nakagawa Y, Hu S, Alverdy J, Kohgo Y, Schneewind O, Jabri B, Chang EB (2007) The *Bacillus subtilis* quorum sensing molecule CSF contributes to intestinal homeostasis via OCTN2, a host cell membrane transporter. Cell Host Microbe 1:299–308

25. Sieuwerts S, de Bok FAM, Hugenholtz J, van Hylckama Vlieg JET (2008) Unraveling microbial interactions in food fermentations: from classical to genomics approaches. Appl Environ Microbiol 74:4997–5007

26. Bolotin A, Quinquis B, Renault P, Sorokin A, Ehrlich SD, Kulakauskas S, Lapidus A, Goltsman E, Mazur M, Pusch GD, Fonstein M, Overbeek R, Kyprides N, Purnelle B, Prozzi D, Ngui K, Masuy D, Hancy F, Burteau S, Boutry M, Delcour J, Goffeau A, Hols P (2004) Complete sequence and comparative genome analysis of the dairy bacterium *Streptococcus thermophilus*. Nat Biotechnol 22:1554–1558

27. Foucaud C, Poolman B (1992) Lactose transport system of *Streptococcus thermophilus*. Functional reconstitution of the protein and characterization of the kinetic mechanism of transport. J Biol Chem 267:22087–22094

28. Hutkins R, Morris HA, McKay LL (1985) Galactokinase activity in *Streptococcus thermophilus*. Appl Environ Microbiol 50:777–780

29. Desmazeaud MJ (1990) Le lait milieu de culture. Microbiol Alim Nutr 8:313–325

30. Hols P, Hancy F, Fontaine L, Grossiord B, Prozzi D, Leblond-Bourget N, Decaris B, Bolotin A, Delorme C, Ehrlich SD, Guédon E, Monnet V, Renault P, Kleerebezem M (2005) New insights in the molecular biology and physiology of *Streptococcus thermophilus* revealed by comparative genomics. FEMS Microbiol Rev 29:435–463

31. Letort C, Nardi M, Garault P, Monnet V, Juillard V (2002) Casein utilization by *Streptococcus thermophilus* results in a diauxic growth in milk. Appl Environ Microbiol 68:3162–3165

32. Shahbal S, Hemme D, Desmazeaud MJ (1991) High cell wall associated proteinase activity of some *Streptococcus thermophilus* strains (H-strains) correlated with a high acidification rate in milk. Lait 71:351–357

33. Herve-Jimenez L, Guillouard I, Guedon E, Gautier C, Boudebbouze S, Hols P, Monnet V, Rul F, Maguin E (2008) Physiology of *Streptococcus thermophilus* during the late stage of milk fermentation with special regard to sulfur amino-acid metabolism. Proteomics 8:4273–4286

34. Withworth DE, Cock PJA (2008) Two-component systems of the myxobacteria: structure, diversity and evolutionary relationships. Microbiology 154:360–372

35. Galperin MY (2005) A census of membrane-bound and intracellular signal transduction proteins in bacteria: bacterial IQ extroverts and introverts. BMC Microbiol 5:35–59

36. Thevenard B, Rasoava N, Fourcassié P, Monnet V (2011) Characterization of *Streptococcus thermophilus* two-component system: in silico analysis, functional analysis and expression of response regulator genes in pure or mixed culture with yogurt partner *Lactobacillus delbrueckii* subsp. *bulgaricus*. Int J Food Microbiol 151:171–181

37. Herve-Jimenez L, Guillouard I, Guedon E, Boudebbouze S, Hols P, Monnet V, Maguin E, Rul F (2009) Postgenomic analysis of *Streptococcus thermophilus* coclutivated in milk with *Lactobacillus delbrueckii* subsp. *bulgaricus*: involvement of nitrogen, purine, and iron metabolism. Appl Environ Microbiol 75:2062–2073

38. Rul F, Ben-Yahia L, Chegdani F, Wrzosek L, Thomas S, Noordine M-L, Gitton C, Cherbuy C, Langella P, Thomas M (2011) Impact of the metabolic activity of *Streptococcus thermophilus* of the colon epithelium of gnotobiotic rats. J Biol Chem 286:10288–10296

39. Williams P, Winzer K, Chan WC, Cámara M (2007) Look who's talking: communication and quorum sensing in the bacterial world. Philos Trans R Soc Lond B Biol Sci 362:1119–1134

40. Rocha-Estrada J, Aceves-Diez AE, Guarneros G, de la Torre M (2010) The RNPP family of quorum sensing proteins in Gram-positive bacteria. Appl Microbiol Biotechnol 87:913–923

41. Fleuchot B, Gitton C, Guillot A, Vidic J, Nicolas P, Besset C, Fontaine L, Hols P, Leblond-Bouget N, Monnet V, Gardan R (2011) Rgg proteins associated with internalized small hydrophobic peptides: a new quorum sensing mechanism in streptococci. Mol Microbiol 80:1102–1119

42. Nychas G-JE, Skandamis P (2005) Fresh meat spoilage and modified atmosphere packaging (MAP). In: Sofos JN (ed) Improving the safety of fresh meat. CRC/Woodhead Publishing Limited, Cambridge, UK, p 461

43. Blana VA, Doulgeraki A, Nychas G-JE (2011) Autoinducer-2–like activity in lactic acid bacteria isolated from minced beef packaged under modified atmospheres. J Food Prot 74:631–635

44. Nychas G-JE, Dourou D, Skandamis P, Koutsoumanis K, Baranyi J, Sofos J (2009) Effect of microbial cell-free meat extract on the growth of spoilage bacteria. J Appl Microbiol 107:1819–1829

45. Pinto UM, Viana ES, Martins ML, Vanetti MCD (2007) Detection of acylated homoserine lactones in gram-negative proteolytic psychrotrophic bacteria isolated from cooled raw milk. Food Control 18:1322–1327

46. Ercolini D, Russo F, Torrieri E, Masi P, Villani F (2006) Changes in the spoilage-related microbiota of beef during refrigerated storage under different packaging conditions. Appl Environ Microbiol 72:4663–4671

47. Jay JM, Vilai JP, Hughes ME (2003) Profile and activity of the bacterial biota of ground beef held from freshness to spoilage at 5–7 °C. Int J Food Microbiol 81:105–111

48. Bruhn JB, Christensen AB, Flodgaard LR, Nielsen KF, Larsen TO, Givskov M, Gram L (2004) Presence of acylated homoserine lactones (AHLs) and AHL-producing bacteria in meat and potential role of AHL in spoilage of meat. Appl Environ Microbiol 70:4293–4302

49. Stanbridge LH, Davies AR (1998) The microbiology of chill-stored meat. In: Davies AR, Board RG (eds) The microbiology of meat and poultry. Blackie Academic & Professional, London, p 174

50. Ferrocino I, Ercolini D, Villani F, Moorhead SM, Griffiths MW (2009) *Pseudomonas fragi* strains isolated from meat do not produce N-Acyl homoserine lactones as signal molecules. J Food Prot 72:2597–2601

51. Widmer KW, Soni KA, Hume ME, Beier RC, Jesudhasan P, Pillai SD (2007) Identification of poultry meat–derived fatty acids functioning as quorum sensing signal inhibitors to autoinducer- 2 (AI-2). J Food Sci 72:M363–M368

52. Lu L, Hume ME, Pillai SD (2004) Autoinducer-2–like activity associated with foods and its interaction with food additives. J Food Prot 67:1457–1462

53. Corry JE, Atabay HI (2001) Poultry as a source of *Campylobacter* and related organisms. Symp Ser Soc Appl Microbiol 30:96S–114S

54. Park SF (2002) The physiology of *Campylobacter* species and its relevance to their role as foodborne pathogens. Int J Food Microbiol 74:177–188

55. Davis MA, Conner DE (2007) Survival of *Campylobacter jejuni* on poultry skin and meat at varying temperatures. Poult Sci 86:765–767

56. El-Shibiny A, Connerton P, Connerton I (2009) Survival at refrigeration and freezing temperatures of *Campylobacter coli* and *Campylobacter jejuni* on chicken skin applied as axenic and mixed inoculums. Int J Food Microbiol 131:197–202

57. Ligowska M, Cohn MT, Stabler RA, Wre BW, Brøndsted L (2011) Effect of chicken meat environment on gene expression of *Campylobacter jejuni* and its relevance to survival in food. Int J Food Microbiol 145:S111–S115

58. Feng P, Weagant SD (2002) Bacteriological analytical manual online. Diarrheagenic *Escherichia coli*. Available from: http://www.cfsan.fda.gov/%7Eebam/bam-4a.html

59. Uhlich GA, Cooke PH, Solomon EB (2006) Analysis of the red-dry-rough phenotype of an *Escherichia coli* O157:H7 strain and its role in biofilm formation and resistance to antibacterial agents. Appl Environ Microbiol 72:2564–2572

60. Hentzer M, Givskov M (2003) Pharmacological inhibition of quorum sensing for the treatment of chronic bacterial infections. J Clin Investig 112:1300–1307

61. Molina L, Constantinescu F, Michel L, Reimmann C, Duffy B, Défago L (2003) Degradation of pathogen quorum sensing molecules by soil bacteria: a preventive and curative biological control mechanism. FEMS Microbiol Ecol 45:71–81

62. Medina-Martìnez MS, Uyttendaele M, Rajkovic A, Nadal P, Debevere J (2007) Degradation of N-acyl-L-homoserine lactones by *Bacillus cereus* in culture media and pork extract. Appl Environ Microbiol 73:2329–2332

63. Fast W, Tipton PA (2012) The enzymes of bacterial census and censorship. Trends Biochem Sci 37:7–14

64. Huang JJ, Han JI, Zhang LH, Leadbetter JR (2003) Utilization of acyl-homoserine lactone quorum signals for growth by a soil pseudomonad and *Pseudomonas aeruginosa* PAO1. Appl Environ Microbiol 69:5941–5949

65. Choo JH, Rukayadi Y, Hwang JK (2006) Inhibition of bacterial quorum sensing by vanilla extract. Lett Appl Microbiol 42:637–641

66. Zhu H, Sun SJ (2008) Inhibition of bacterial quorum sensing-regulated behaviors by *Tremella fuciformis* extract. Curr Microbiol 57:418–422

67. Givskov M, de Nys R, Manefield M, Gram L, Maximilien R, Eberl L, Molin S, Steinberg PD, Kjelleberg S (1996) Eukaryotic interference with homoserine lactone-mediated prokaryotic signaling. J Bacteriol 178:6618–6622

68. Vasanthi HR, Mukherjee S, Das DK (2009) Potential health benefits of broccoli – a chemico-biological overview. Mini Rev Med Chem 9:749–759

69. Sperandio V, Torres AG, Jarvis B, Nataro JP, Kaper JB (2003) Bacteria–host communication: the language of hormones. Proc Natl Acad Sci USA 100:8951–8956

70. Jakobsen TH, Bragason SK, Phipps RK, Christensen LD, van Gennip M, Alhede M, Skindersoe M, Larsen TO, Høiby N, Bjarnsholt T, Givskov M (2012) Food as a source for quorum sensing inhibitors: iberin from horseradish revealed as a quorum sensing inhibitor of *Pseudomonas aeruginosa*. Appl Environ Microbiol 78:2410–2421

71. Wang R, Starkey M, Hazan R, Rahme LG (2012) Honey's ability to counter bacterial infections arises from both bactericidal compounds and QS inhibition. Front Microbiol 3:1–8

Chapter 4
The Probiotic Message

4.1 Introduction

Probiotic organisms are live microorganisms that, when administrated in adequate amounts, confer a health benefit to the host [1]. This definition implies that two conditions have to be satisfied: close interaction between probiotic and host, and microbial adaptation to the selective host environment. One of the main health-promoting effects exerted by probiotics is in relation to the inhibition of pathogens at the human gastrointestinal level. This inhibitory activity mainly relies on competition for nutrients, cooperation for nutrients with other beneficial species, synthesis of antimicrobial compounds and competitive exclusion. The promotion of symbiotic and syntrophic relationships with other beneficial intestinal species is stimulated through the formation of biofilm, and bacteriocins are, probably, some of the most effective antibacterial compounds in this ecosystem. Both these phenotypes may rely on quorum sensing mediated control (see Sects. 2.3 and 2.4). Other sophisticated mechanisms, which are mediated by probiotics, also condition the interactions between microbiota and host. They include the restoration of microbial homeostasis through microbe-microbe interactions, the enhancement of epithelial barrier function, and the modulation of immune responses [2–4]. Also under these circumstances, various mechanisms of communication may occur and assume the role of coordinating the adaptation processes, either within the microbial community or between probiotics and host.

Probiotics are delivered by food or pharmaceutical preparations, and at least in some aspects they have an impact on human health. This chapter focuses on the most relevant evidence relating to the bacterial quorum sensing mechanisms that occur at gastrointestinal level between probiotic bacteria and intestinal inhabitants, and between probiotics and host.

M. Gobbetti and R. Di Cagno, *Bacterial Communication in Foods*,
SpringerBriefs in Food, Health, and Nutrition, DOI 10.1007/978-1-4614-5656-8_4,
© Marco Gobbetti and Raffaella Di Cagno 2013

4.2   Language-Like Exchange at Intra- and Interspecies Levels

During the last decade, the molecular languages used for exchange within and among microbial species in the gut environment have become better understood. At the gastrointestinal level, symbiotic bacteria may synthesize, detect, and respond to several signaling molecules, which have low molecular masses and various chemical structures (e.g., lacton-like autoinducers, peptide pheromones, and autoinducer-2, including furanones) (see Chap. 1). Cross-talk at intra- and interspecies levels is based on the exchange of these signaling molecules from both directions. During this exchange, a physiological effect on the human host may occur, and this may vary depending on the signal molecule.

Overall, quorum sensing molecules display broad biological activities, which are well beyond their specific role in bacterial communication. This is the case for N-acyl-L-homoserine lactones (AHL) synthesized by *Pseudomonas aeruginosa*, which exhibit antibacterial, pharmacological, and immune modulatory activities [5]. Secretion of proteins across biological membranes is the crucial mechanism by which bacteria may interact themselves and monitor the gastrointestinal environment [6]. The signal molecule autoinducer-2 (AI-2) and its cognate synthase LuxS are considered to be attractive candidates for multispecies communication at the gastrointestinal level [7]. Some probiotic strains of *Lactobacillus rhamnosus*, *L. plantarum*, *L. johnsonii*, and *L. casei*, which very often colonize the intestine, share the capacity to synthesize AI-2-type molecules [8]. The accumulation of two signaling molecules, cholerae AI-1 (CAI-1) and AI-2, repressed the synthesis of virulence factors by *Vibrio cholerae*, when high population density is reached. AI-2 acts in a synergistic way with CAI-1 and controls the expression of virulence genes [9]. AI-2 was also found to be responsible for the probiotic activity of *L. reuteri*, an autochthonous inhabitant of the rodent forestomach. In this case, the functional role of AI-2 is to control the formation of biofilm, which permits the adhesion of the bacterium to nonsecretory epithelium. The disruption of the *luxS* gene modified the capacity of *L. reuteri* to form biofilm either in vitro or on the epithelial surface of murine forestomach [10]. Compared to wild type, the *luxS* mutant showed thicker biofilm and decreased levels of ATP during the exponential phase of growth [10, 11]. The ecological performance of the *luxS* mutant was significantly decreased, especially under the highly competitive conditions of the murine cecum [10]. The microarray comparison between the profiles of gene expression of *L. reuteri* wild type and *luxS* mutant, revealed the altered transcription of genes, which encode for proteins associated with cysteine biosynthesis/oxidative stress response, urease activity, and sortase-dependent proteins [11]. As shown through metabolomic analyses, the *luxS* mutation also affected the levels of fermentation end products, fatty acids and amino acids. Another strain of *L. reuteri* showed the capacity to alter the virulence of *Staphylococcus aureus* via secretion of cell-to-cell signaling molecules [12]. Under co-culture conditions, this strain synthesized one or more molecules that inhibit the expression of superantigen-like protein 11 (SSL11), putative staphylococcal exotoxin and RNAIII, the effector molecule of the staphylococcal *agr* locus.

The disruption of the *luxS* gene of *L. rhamnosus* GG resulted in pleiotropic effects, which were attributed to metabolic defects. In particular, they concerned the rate of growth, biofilm formation, and the in vivo persistence on the murine gastrointestinal tract [13, 14].

Chronologically, bifidobacteria are among the bacterial groups that first colonize the gastrointestinal tract. These bacteria are thought to exert several health-promoting effects. Probably, the most important are the modulation of the immune response and the maintenance of intestinal barrier integrity [15]. Recently, factors that are involved in the bifidobacterial cross-talk emerged. It seems that LuxS plays a key role in the quorum sensing regulation of *Bifidobacterium longum* [16]. The physiology of *B. longum* and *B. breve* was investigated under co-culture conditions [17]. Compared to mono-culture conditions, a shift of some enzyme activities that are responsible for the degradation of complex carbohydrates (e.g., β-D-xylopyranosidase, α-L-arabinofuranosidase) was observed [17]. This effect was attributed to various causes such as the diverse substrate specificity between the species, nutrient competition, and enzyme induction by available carbohydrates [18]. The proteomic analysis strengthened the role of quorum sensing mechanisms during co-cultivation. Sixteen proteins varied their levels of expression and, in particular, ten ribosomal proteins were up-regulated in *B. longum* and *B. breve*. Ribosomal proteins are necessary for ribosome assembly and stability, and in certain bacteria they are implicated in sensing environmental changes [19]. The transcriptional regulator (ClgR), which is involved in the regulation of the *clpC* gene and *clpP* operon [20], was up-regulated in *B. breve*. ClpC belongs to the stress-response-related Clp ATPase family, whose members act as chaperones and regulators of proteolysis. Protein products of the *clpP* operon display significant homology to the characterized ClpP peptidases [21]. The *ClpC* gene and *clpP* operon are involved in the response of *B. breve* to some environmental stimuli (e.g., heat treatment). The same up-regulation was found for glycosyltransferase, which is involved in the biosynthesis of peptidoglycan and cell division [22]. Probably, the enhanced cell wall biosynthesis by *B. breve* represents the mechanism of response to the inhibition exerted by *B. longum*. Three different proteolytic products of the enzyme Xfp (fructose 6-phosphate phosphoketolase) were up-regulated in *B. longum*. The general interpretation of the above results was that strains sense each other and rearrange their carbohydrate metabolism to increase their capacity to compete.

The adaptation of *B. longum* to the gastrointestinal tract was reflected by proteomic changes both under in vitro and in vivo incubation [16]. Under conditions that mimicked the intestinal growth, 14 proteins were up-regulated and four proteins showed changes in mobility. Most of these proteins were related to stress response and translation. For instance, variations of the level of expression were found for the elongation factor Tu (EF-Tu), which contributes to retention or attachment, and acts as a *Bifidobacterium* adhesion-like factor. Other variations were found for bile salt hydrolase (BSH), which promotes interaction between probiotics and the intestine, and, more generally, for stress proteins, which defend the bacterium from bile salts and other harmful compounds at the gastrointestinal level. Four proteins such as glutamine synthetase (GlnA1), phosphoribosylaminoimidazole-succinocarboxamide synthase

(PurC), LuxS and phosphoglycerate kinase (Pgk) exhibited a clear post-translational modification. Western blot analysis and the use of phosphospecific ProQ-Diamond stain revealed that substances from the gastrointestinal tract trigger the phosphorylation of Pgk and LuxS at the Ser/Thr residues. For the first time, these proteins were identified as bifidobacterial phosphoproteins. The phosphorylation of LuxS may represent the mechanism that enteric bacteria evolved to interfere with the signaling capabilities of neighboring bacterial species. In particular, two different mechanisms of interference were proposed for *B. longum*: (1) the use of the signals released from neighboring bacteria to recognize the presence of AI-2 within the intestine, especially when large numbers of bacteria are present; and/or (2) the response to competitors by sequestering and destroying foreign AI-2, thus disturbing intercellular communication.

The secretome of *B. longum* was also studied to obtain information on the monitoring of the local environment [23]. Seventeen proteins were detected and some of them were identified. They mainly included two hypothetical solute-binding proteins of ABC transporters for peptides, the phosphate-binding transport protein of the ABC transport system, cell wall synthetase, cell division-specific transpeptidase responsible for septum formation, two cell wall-associated hydrolases, putative enzymes catalyzing cell wall turnover, and several polypeptides with similarity to bacterial conjugation proteins. All the above proteins were predicted to contain a signal peptide, and a signal peptide cleavage site in their immature form.

A further example of signal exchange involves *L. acidophilus*, a well-known probiotic bacterium, which is widely used for the manufacture of fermented milk products [24]. Co-cultivation of this species with the yoghurt starters, *Streptococcus thermophilus* and *L. delbrueckii* subsp. *bulgaricus*, induced the synthesis of the bacteriocin lactacin B by *L. acidophilus*. The induction was not found when heat-killed cells or their cell-free supernatants were used. The induction of lactacin B was only promoted when live cells of yogurt starters were used for co-cultivation. The cell-to-cell contact seemed to be indispensable. Probably, a cell wall-associated factor induced the synthesis of the signal molecule. The autoinduction of the synthesis of lactacin B responds only when target cells are alive and coexist in the same environment with the bacteriocin producer.

4.3   Language-Like Exchange at Interkingdom Level

Prokaryotes and eukaryotes have coexisted for millions of years. It is estimated that humans have 10^{13} human cells and 10^{14} bacterial cells (comprising the endogenous microbiota). Eukaryotes have a variable relationship with prokaryotes, and these interactions may be either beneficial or detrimental. Humans maintain a symbiotic association with their intestinal microbiota, which is crucial for nutrient assimilation and development of the innate immune system [25]. Recently, it was discovered that bacteria might also exploit quorum sensing signals to communicate with the host under a process defined as cross-kingdom cell-to-cell signaling [26].

This communication involves small molecules such as hormones, which are produced by eukaryotes, and hormone-like chemicals, which are synthesized by bacteria. These signals, however, may be hijacked by bacterial pathogens to activate their virulence genes. Mammalian hormones include proteins or peptides, steroids (a subclass of lipidic hormones), and amino acid derivatives or amines. The structure of the hormone dictates the location of its receptor. Amine and peptide hormones cannot cross the cell membrane and bind to cell surface receptors (e.g., receptor kinases and G-protein coupled receptors). Steroid hormones may cross plasma membranes and primarily bind to intracellular receptors. Protein and peptide constitute most of the hormones; they contain 3–200 amino acids and are usually post-translationally processed.

Several of the mechanisms used for hormonal communication between bacteria and their hosts refer to pathogenic interactions [26]. Bacteria sense and respond to amine hormones (adrenaline and noradrenaline) to regulate a multitude of phenotypes that range from metabolism to virulence gene expression. Noradrenaline stimulates the growth, and induces the synthesis of fimbriae and toxins in pathogenic strains of *Escherichia coli* [27–29]. Both these hormones induce the expression of flagella and a type III secretion system (T3SS) in enterohemorrhagic *E. coli* O157:H7. This strain may sense either the host adrenaline and noradrenaline or the bacterial aromatic quorum sensing signal AI-3 (see Sect. 1.6) to express its virulence traits. This suggested that the above signals are interchangeable [30] (Fig. 4.1). The histidine sensor kinase QseC, which is located in the cell membrane of *E. coli* O157:H7, senses the signals, which results in its auto-phosphorylation. QseC phosphorylates its response regulator QseB and initiates a phosphorelay signaling cascade that activates the expression of a second two-component system (QseEF), The cascade culminates with the activation of several genes. These correspond to *flhDC*, which are responsible for motility, *ler* (locus of enterocyte effacement [LEE]-encoded regulator), which encode the components of a type III section system involved in attaching and effacing (AE) lesion formation, and *stxAB*, which are responsible for Shiga-toxin expression [26]. Recently, it was shown that lactobacilli interfered with this mechanism of signaling. The supernatant of *L. reuteri* exhibited a negative effect on the expression of the *ler* gene of *E. coli* O157:H7 [14]. Similarly, *L. acidophilus* decreased the expression of the virulence genes [14]. The LuxS-dependent production of an unknown competing antagonistic molecule by *L. reuteri* and *L. acidophilus* was suggested [14]. The main target of this molecule is the AI-3-like agonistic compound. On the contrary, the expression of the *ler* gene was induced by the supernatant of the stationary phase culture of another strain of *L. reuteri*. The induction was abolished in the presence of the isogenic *luxS* mutant. Under these circumstances, the synthesis of the AI-3-like agonistic molecule was hypothesized [14].

Probiotic lactobacilli are suggested to strengthen the human epithelial barrier. Various mechanisms were suggested for this activity: induction of mucin secretion, enhancement of tight-junction functioning, up-regulation of cytoprotective heat shock proteins, and prevention of the apoptosis of epithelial cells [14]. Although limited attention was paid to interactions between quorum sensing molecules (e.g.,

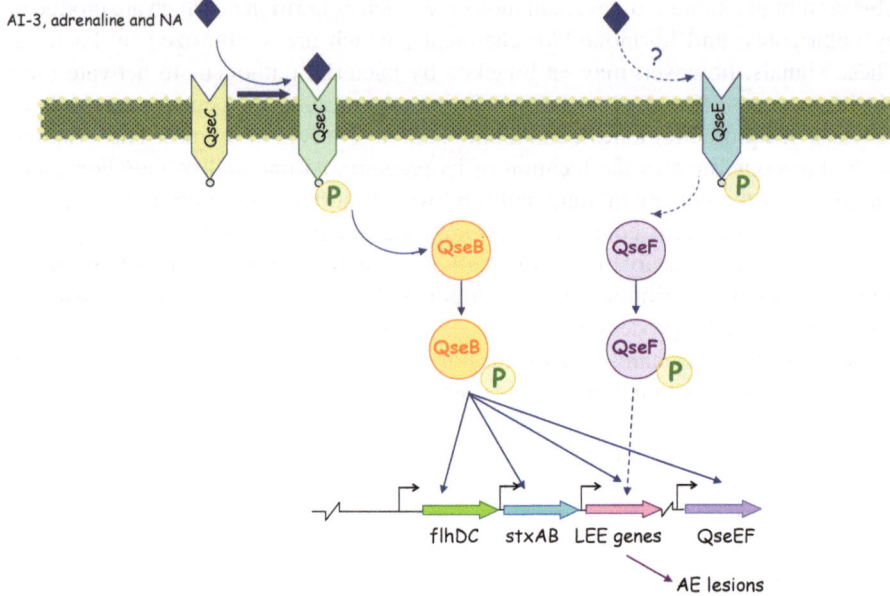

Fig. 4.1 Example of hormonal communication in enterohemorrhagic *Escherichia coli* O157:H7. Autoinducer (AI)-3, adrenaline and noradrenaline (NA) bind the bacterial receptor sensor kinase QseC, which results in its auto-phosphorylation and a complex regulatory cascade is activated. QseB, response regulator; QseEF, two-component system; *flhDC*, motility gene; *stxAB*, Shiga toxin gene; *ler*, locus of enterocyte effacement (*LEE*)-encoded regulator; AE, attaching and effacing lesion

peptide pheromones) and human intestinal mucosa, some features of these signaling pathways were identified (Fig. 4.2). The quorum sensing pentapeptide (sequence ERGMT), known as the competence and sporulation factor (CSF), of probiotic *Bacillus subtilis* activated key survival pathways at intestinal level. The activation was mainly observed for p38 MAP kinase, protein kinase B (Akt), and cytoprotective heat shock proteins, which prevented oxidative intestinal cell injury and loss of the barrier function [31]. The effect of CSF depends on the uptake via the apical membrane organic cation transporter-2 (OCTN2). This transport is an example of host bacterial interaction, which allows the host to monitor and respond to changes in the behavior or composition of the colonic microbiota (Fig. 4.3). OCTN2 and similar pathways to engage or uptake quorum sensing molecules may be essential to regulate the host response and to maintain intestinal homeostasis. A probiotic effect and high survival during gastrointestinal transit were diffusely reported for strains of *L. plantarum*. These features allow one to consider this species as a promising candidate for human health-related delivery of functional molecules [32]. Language-like exchange at the interkingdom level was found between pheromone PlnA, which is synthesized by sourdough *L. plantarum*, and human Caco-2/TC7 cells (human colon carcinoma) [33] (see Sect. 3.2.2). In particular, PlnA increased the viability of Caco-2/TC7 cells, which represent one of the in vitro systems most

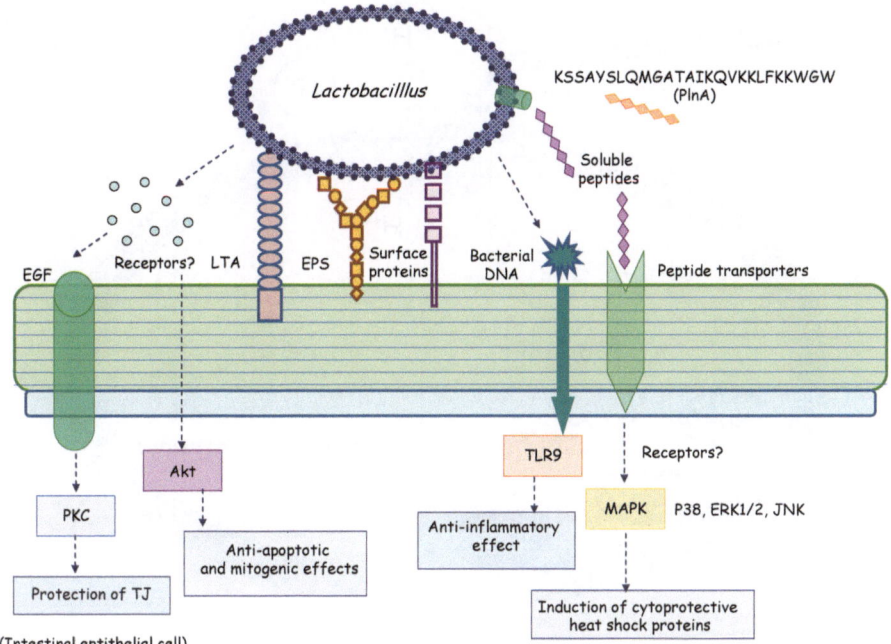

Fig. 4.2 Example of modulation of the intestinal epithelial barrier function by lactobacilli. Signaling pathways involved in the interaction between lactobacilli and human intestinal cells were identified [14]. Direct contact between lactobacilli and epithelial cells by exopolysaccharides (*EPS*), surface proteins, and lipoteichoic acid molecules (*LTA*) are needed for beneficial effects. MAP kinase p38, ERK1/2, and JNK (*MAPK*) have a key role in the dynamic regulation of the cell cytoskeleton, tight junctions (*TJ*), and other effectors of epithelium barrier function. The epidermal growth factor (*EGF*) receptor is another signaling pathway, which may be activated by soluble proteins secreted by lactobacilli and, in turn, it activates the protein kinase B (*Akt*). This signaling is responsible for promoting the survival of epithelial cells through the inactivation of several anti-apoptotic pathways. Bacterial DNA via Toll-like receptor 9 (*TLR9*) signaling may have anti-inflammatory effects. Pheromone plantaricin A (*PlnA*) synthesized by sourdough *Lactobacillus plantarum* DC400 prevented the damage effect of interferon-γ towards Caco-2/TC7 cells and transepithelial electrical resistance [33]

widely used to mimic intestinal mucosa [34]. Compared to the negative control (medium alone), biologically or chemically synthesized PlnA markedly increased the level of transepithelial electric resistance (TEER). PlnA also prevented the damage caused by interferon-γ towards Caco-2/TC7 cells and TEER, thus eliminating the negative effect of cytokines. The PlnA contained in the cell-free supernatant of the co-culture between sourdough *L. plantarum* DC400 and *L. sanfranciscensis* DPPMA174 was studied for its protective effect on human NCTC 2544 keratinocytes [35]. Overall, the skin acts as a physical barrier, exerting fluid homeostasis, thermoregulation, and protection against oxidative stresses and infections [36, 37]. Keratinocytes are the primary sensors of invading pathogens through recognition of various evolutionarily conserved microbial components. PlnA protected keratinocytes

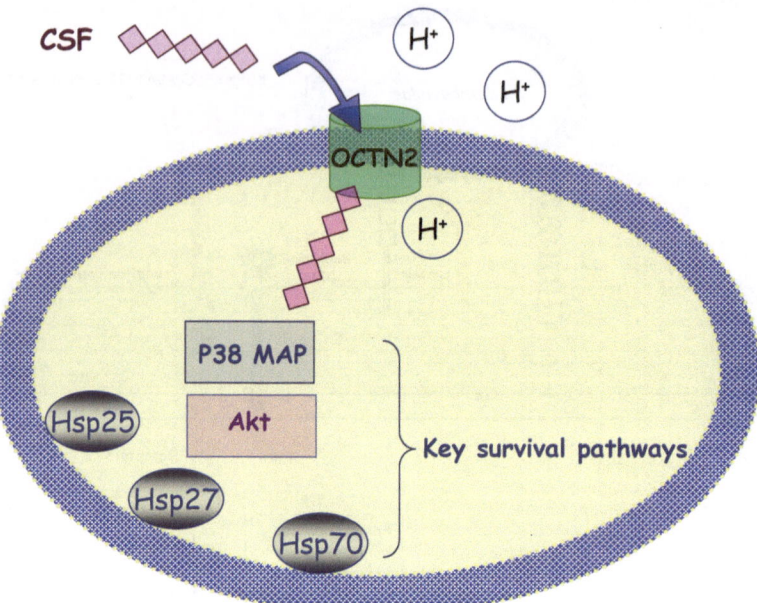

Fig. 4.3 Probiotic message signaling competence and sporulation factor (*CSF*) spoken by the Gram-positive bacterium *Bacillus subtilis* transposed in the host via an organic cation transporter-2 (*OCTN2*). Heat shock proteins, *HSP*, protein kinase B, *Akt*

against oxidative stress and modulated the expression of filaggrin (FLG), involucrin (IVL), β-defensin 2, and tumor necrosis factor-α (TNF-α) genes. Protection against oxidative stress was attributed to various factors such as the increased keratinocyte viability [38], the decreased generation of intracellular ROS (reactive oxygen species) [37], and direct or indirect enhancement of the antioxidant system. The primary structure of PlnA includes several amino acids (e.g., Lys, Ala, Tyr, Leu, Met, Gly, Trp, and Val), which may impart a certain degree of antioxidative potential [39]. These findings suggested that the peptide pheromone PlnA was positively sensed by human keratinocytes. The health-promoting effect of this language/signal (PlnA) on tissue regeneration may encourage its use for dermatological purposes.

References

1. FAO/WHO (2001) Evaluation of health and nutritional properties of powder milk and live lactic acid bacteria. Food and Agriculture Organization of the United Nations and World Health Organization expert consultation report, FAO, Rome
2. Boirivant M, Strober W (2007) The mechanism of action of probiotics. Curr Opin Gastroenterol 23:679–692
3. Fedorak RN, Madsen JL (2004) Probiotics and the management of inflammatory bowel disease. Inflamm Bowel Dis 10:286–299

4. Marco ML, Pavan S, Kleerebezem M (2006) Towards understanding molecular modes of probiotic action. Curr Opin Biotechnol 17:204–210
5. Diggle SP, Matthijs S, Wright VJ, Fletcher MP, Chabra SR, Lamont IL, Kong X, Hider RC, Cornelis P, Cámara M, Williams P (2007) The *Pseudomonas aeruginosa* 4-quinolone signal molecules HHQ and PQS play multifunctional roles in quorum sensing and iron entrapment. Chem Biol 14:87–96
6. Van Pijkeren JP, Canchaya C, Ryan KA, Li Y, Claesson MJ, Sheil B, Steidler L, O'Mahony L, Fitzgerald GF, van Sinderen D, O'Toole PW (2006) Comparative and functional analysis of sortase-dependent proteins in the predicted secretome of *Lactobacillus salivarius* UCC118. Appl Environ Microbiol 72:4143–4153
7. Lebeer S, Vanderleyden J, De Keersmaecker SCJ (2008) Genes and molecules of lactobacilli supporting probiotic action. Microbiol Mol Biol Rev 72:728–764
8. De Keersmaecker SCJ, Vanderleyden J (2003) Constraints on detection of autoinducer-2 (AI-2) signalling molecules using *Vibrio harvey* as a reporter. Microbiology 149:1953–1956
9. Higgins DA, Pomianek ME, Kraml CM, Taylor RK, Semmelhack MF, Bassler BL (2007) The major *Vibrio cholarae* autoinducer and its role in virulence factor production. Nature 450:883–886
10. Tannock GW, Ghazally S, Walter J, Loach D, Brooks H, Cook G, Surette M, Simmers C, Bremer P, Dal Bello F, Hertel C (2005) Ecological behaviour of *Lactobacillus reuteri* 100–23 is affected by mutation of the *luxS* gene. Appl Environ Microbiol 71:8419–8425
11. Wilson CM, Aggio RBM, O'Toole PW, Villas-Boas S, Tannock GW (2012) Transcriptional and metabolomic consequences of *luxS* inactivation reveal a metabolic rather than quorum sensing role for LuxS in *Lactobacillus reuteri* 100–23. J Bacteriol 194:1743–1746
12. Laughton JM, Devillard E, Heinrichs DE, Reid G, McCormick JK (2006) Inhibition of expression of a staphylococcal superantigen-like protein by a soluble factor from *Lactobacillus reuteri*. Microbiology 152:1155–1167
13. Lebeer S, Claes IJJ, Verhoeven TLA, Shen C, Lambrichts I, Ceuppens JL, Vanderleyden J, De Keersmaecker SCJ (2008) Impact of *luxS* and suppressor mutations on the gastrointestinal transit of *Lactobacillus rhamnosus* GG. Appl Environ Microbiol 74:4711–4718
14. Lebeer S, De Keersmaecker SCJ, Verhoeven TLA, Fadda AA, Marchal K, Vanderleyden J (2007) Functional analysis of *luxS* in the probiotic strain *Lactobacillus rhamnosus* GG reveals a central metabolic role important for growth and biofilm formation. J Bacteriol 189:860–871
15. Leahy SC, Higgins DG, Fitzgerald GF, van Sinderen D (2005) Getting better with bifidobacteria. J Appl Microbiol 98:1303–1315
16. Yuan J, Wang B, Sun Z, Bo X, Yuan X, He X, Zhao H, Du X, Wang F, Jiang Z, Zhang L, Jia L, Wang Y, Wei KH, Wang J, Zhang X, Sun Y, Huang L, Zeng M (2008) Analysis of host-inducing proteome changes in *Bifidobacterium longum* NCC2705 grown in vivo. J Prot Res 7:375–385
17. Ruiz L, Sánchez B, de los Reyes-Gávilan CG, Gueimond M, Margolles A (2009) Coculture of *Bifidobacterium longum* and *Bifidobacterium breve* alters their protein expression profiles and anzymatic activities. Int J Food Microbiol 133:148–153
18. Gueimonde M, Noriega L, Margolles A, de los Reyes-Gavilán CG (2007) Induction of alpha-L-arabinofuranosidase activity by monomeric carbohydrates in *Bifidobacterium longum* and ubiquity of encoding genes. Arch Microbiol 187:145–153
19. Chen H, Teplitski M, Robinson JB, Rolfe BG, Bauer WD (2003) Proteomic analysis of wild-type *Sinorhizobium meliloti* responses to N-acyl homoserine lactone quorum sensing signal and the transition to stationary phase. J Bacteriol 185:5029–5036
20. De Dea Lindner J, Canchaya C, Zhang Z, Neviani E, Fitzgerald GF, van Sinderen D, Ventura M (2007) Exploting *Bifidobacterium* genomes: the molecular basis of stress response. Int J Food Microbiol 120:13–24
21. Ventura M, Fitzgerald GF, van Sinderen D (2005) Genetic and transcriptional organization of the clpC locus in *Bifidobacterium breve* UCC 2003. Appl Environ Microbiol 71:6282–6291
22. Mohammadi T, Karczmarek A, Crouvoisier M, Bouhss A, Mengin-Lecreulx D, den Blaauwen T (2007) The essential peptidoglycan glycosyltransferase MurG forms a complex with proteins

involved in lateral envelop growth as well as with proteins involved in cell division in *Escherichia coli*. Mol Microbiol 65:1106–1121

23. Sánchez B, Champomier-Vergès MC, Anglede P, Baraige F, de los Reyes-Gavilàn CG, Margolles A, Zagorec M (2008) A preliminary analysis of *Bifidobacterium longum* exported proteins by two-dimensional electrophoresis. J Mol Microbiol Biotechnol 14:74–79

24. Tabasco R, Paarup T, Janer C, Pelaez C, Requena T (2007) Selective enumeration and identification of mixed cultures of *Streptococcus thermophilus*, *Lactobacillus delbrueckii* subsp. *bulgaricus*, *L. acidophilus*, *L. paracasei* subsp. *paracasei* and *Bifidobacterium lactis* in fermented milk. Int Dairy J 17:1107–1114

25. Hooper LV, Gordon JI (2001) Commensal host–bacterial relationships in the gut. Science 292:1115–1118

26. Hughes DT, Sperandio V (2008) Inter-kingdom signalling: communication between bacteria and their hosts. Nat Rev Microbiol 6:111–120

27. Freestone PP, Haigh RD, Williams PH, Lyte M (2003) Involvement of enterobactin in norepi-nephrine mediated iron supply from transferrin to enterohaemorrhagic *Escherichia coli*. FEMS Microbiol Lett 222:39–43

28. Freestone PP, Lyte M, Neal CP, Maggs AF, Haigh RD, Williams PH (2000) The mammalian neuroendocrine hormone norepinephrine supplies iron for bacterial growth in the presence of transferrin or lactoferrin. J Bacteriol 182:6091–6098

29. Burton CL, Chhabra SR, Swift S, Baldwin TJ, Withers H, Hill SJ, Williams P (2002) The growth response of *Escherichia coli* to neurotransmitters and related catecholamine drugs requires a functional enterobactin biosynthesis and uptake system. Infect Immun 70:5913–5923

30. Sperandio V, Torres AG, Jarvis B, Nataro JP, Kaper JB (2003) Bacteria–host communication: the language of hormones. Proc Natl Acad Sci USA 100:8951–8956

31. Fujiya M, Musch MW, Nakagawa Y, Hu S, Alverdy J, Kohgo Y, Schneewind O, Jabri B, Chang EB (2007) The *Bacillus subtilis* quorum sensing molecule CSF contributes to intestinal homeo-stasis via OCTN2, a host cell membrane transporter. Cell Host Microbe 1:299–308

32. Diep DB, Mathiesen G, Eijsink VG, Nes IF (2009) Use of lactobacilli and their pheromone-based regulatory mechanism in gene expression and drug delivery. Curr Pharm Biotechnol 10:62–73

33. Di Cagno R, De Angelis M, Calasso M, Vincentini O, Vernocchi P, Ndagijimana M, De Vincenzi M, Dessi MR, Guerzoni ME, Gobbetti M (2010) Quorum sensing in sourdough *Lactobacillus plantarum* DC400: induction of plantaricin A (PlnA) under co-cultivation with other lactic acid bacteria and effect of PlnA on bacterial and Caco-2 cells. Proteomics 10:2175–2190

34. Yan F, Cao H, Cover TL, Whitehead R, Washington MK, Polk DB (2007) Soluble proteins produced by probiotic bacteria regulate intestinal epithelial cell survival and growth. Gastroenterology 132:562–575

35. Marzani B, Pinto D, Minervini F, Calasso M, Di Cagno R, Giuliani G, Gobbetti M, De Angelis M (2012) The antimicrobial peptide pheromone Plantaricin A increases antioxidant defenses of human keratinocytes and modulates the expression of filaggrin, involucrin, β-defensin 2 and tumor necrosis factor-α genes. Exp Dermatol. doi:10.1111/j.1600-0625.2012.01538.x

36. Morini F, Dusatti F, Bonina FP, Saija A, Ferro M (2000) Iron induced lipid peroxidation in human skin-derived cell lines: protection by a red orange extract. ATLA Altern Lab Anim 28:427–433

37. Calcabrini C, De Bellis R, Mancini U, Cucchiarini L, Potenza L, De Sanctis R, Patrone V, Scesa C, Dachà M (2010) Rhodiola rosea ability to enrich cellular antioxidant defences of cultured human keratinocytes. Arch Dermatol Res 302:191–200

38. Pinto D, Marzani B, Minervini F, Calasso M, Giuliani GM, Gobbetti M, De Angelis M, Plantaricin A (2011) synthesized by *Lactobacillus plantarum* induces in vitro proliferation and migration of human keratinocytes and increases the expression of TGF-β1, FGF7, VEGF-A and IL-8 genes. Peptides 32:1815–1824

39. Sarmadi BH, Ismail A (2010) Antioxidative peptides from food proteins: a review. Peptides 31:1949–1956

Chapter 5
The New Perspective

5.1 Introduction

An abundant literature has indubitably established that cross-talk between bacteria occurs and variously influences bacterial phenotype, environmental adaptation and behavior under a multitude of circumstances. Although the majority of the examples provided in the literature concerned phyto- and human pathogenic bacteria, bacterial communication in foods is emerging. Understanding the communication between food-related or health-promoting bacteria may allow one to: (1) select new antimicrobial compounds mainly based on quorum quenching mechanisms; (2) select starters more efficiently and decrease the risk of fermentation failure; (3) enhance the hygiene, sensory, nutritional, and shelf-life properties of foods; and (4) develop novel biogenic compounds.

5.2 Antimicrobial Therapy

The emergence of antibiotic-resistant bacteria poses a global threat to human health and it is classified as the clinical super challenge of the twenty-first century [1]. Despite the continued emergence of antibiotic-resistant strains, the development of new and effective antibiotic treatments seems to be inadequate. Compared to classical treatments, alternative antimicrobial therapies have to emerge. Therapies that target pathways of bacterial quorum sensing are promising. The inhibition of quorum sensing is potentially advantageous, not only as it removes the selective pressure for resistance mechanisms but, especially, as it controls the bacterial virulence factors that are responsible for human infections. The large number of receptors, transporters, regulators, and signals represent multiple targets to switch off quorum sensing circuits. For instance, the autoinducer-2 (AI-2) quorum sensing signal has already been established as one of the main causes of bacterial pathogenicity in humans (see Sect. 2.2). The ubiquitous nature of AI-2 makes it an excellent target for quorum quenching and,

M. Gobbetti and R. Di Cagno, *Bacterial Communication in Foods*,
SpringerBriefs in Food, Health, and Nutrition, DOI 10.1007/978-1-4614-5656-8_5,
© Marco Gobbetti and Raffaella Di Cagno 2013

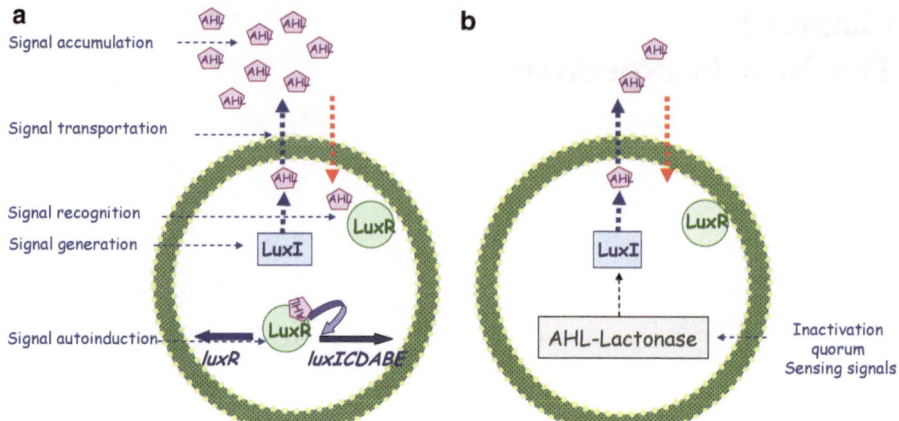

Fig. 5.1 Schematic representation of N-acyl-L-homoserine lactone (*AHL*)-dependent quorum sensing (**a**) and quorum quenching (**b**) in Gram-negative bacteria. Key processes that could be targeted by quorum quenching through inactivation (e.g., by plasmid curing or expressing the lactonase) of AHL signals are shown (Adapted from [3])

consequently, a potential antimicrobial therapy. Because AI-2 is not fundamental for cell growth or survival, interference with the synthesis and processing of this molecule will not promote microbial resistance. Nevertheless, because of the high degree of diversity among bacterial species, the most promising approaches have to include a cocktail of quorum sensing quenchers and traditional antibiotics. Strains of *Pseudomonas aeruginosa*, which were treated with garlic extract as a quorum quencher, were also more sensitive to the antibiotic tobramycin [2].

Overall, the mechanisms of quorum quenching may block generation, disturb exchange, prevent recognition, and trap and inactivate quorum sensing signals [3] (Fig. 5.1). In particular, N-acyl homoserine lactone (AHL)-degrading enzymes with activity that works as a type of censorship to block interbacterial communication may represent useful tools to condition microbial ecology and to increase our knowledge on enzymes that underlie the social lives of bacteria [4]. In nature, whenever a bacterium evolves a competitive advantage, it is almost an inevitable corollary that competing bacteria will develop interfering strategies. The best-known quorum quenching enzymes are categorized into two distinct groups: AHL lactonases and AHL acylases. Although the role of these enzymes in their native environments has to be elucidated in more depth, their utility as biochemical tools as well as in potential industrial and therapeutic applications is quite promising [4].

5.3   Selection of Starters

Food and beverage fermentations are typically carried out with mixed culture starters, which consist of multiple strains or species having primary economic importance. Indeed, the fermentation is not the simple result of adding up individual strain functionalities but it is the consequence of microbial interactions, which take place at the level of substrates, and are mediated through exchange of metabolites, growth factors, or inhibiting compounds. Because of the heterogeneous physicochemical composition of the food matrices it is generally the case that one finds simultaneous occupation of multiple niches by specialized strains. Furthermore, the food matrix represents a source of autochthonous microorganisms, which associate with or out-compete the starter cultures [5–9]. The sum of these interactions affects the dynamics and behavior of the resulting mixed population. On the one hand, bacteria, as members of a mixed population, have to evolve and compete through optimization of their fitness. This is achieved via specialization, for instance through the optimization of specific metabolisms that are the most suitable for a certain environment. On the other hand, the priority in food fermentation is to obtain stable and performing microbial consortia over time. Although impressive results were obtained from genetic and proteomic points of view, which allowed controlled fermentation processes, the discovery of cell-to-cell communication opens up a new era on how such microbial populations are perceived and controlled. Several questions have to be solved to investigate such aspects in greater depth. For instance, can the technological processes and characteristics of the complex food matrices influence the bacterial signaling pathways? How many signaling molecules are generated during food fermentation and how many signals are sensed by bacteria? How variable is the chemical structure of the signals? And, how many receptors do the bacteria possess?

Certainly, the criteria to select starters for food fermentation have to combine pro-technological features, functionality, and robustness with the monitoring of their capacity to communicate. Progress in this regard was initially made on sourdough and yoghurt starter cultures (see Chap. 3), but the perspective needs to be enlarged to encompass other food ecosystems. Regarding Gram-positive bacteria, the selection and use of starter cultures, which have the capacity to synthesize small antimicrobial peptides (AMP), should be carefully considered. This should be aimed not only at considering the specific spectrum of activity but also at investigating the constitutive versus regulated nature of the AMP, how this regulation takes place and whether or not the synthesis of AMP significantly enhances starter performance within a specific food ecosystem.

5.4   Food Quality

Evidence was reported of a direct correlation between the presence of signaling compounds (e.g., AHL and AI-2) and the contamination of foods by specific or ephemeral spoilage microorganisms [10]. Bacterial phenotypes, which are regulated via AHL, may influence the sensory, nutritional, and hygiene features of several foods (see Sect. 3.4.1). Overall, it seems that the synthesis of AI-2-like molecules affects the dominance of various bacterial strains during food storage, especially interfering with their persistence. Nevertheless, data regarding quorum sensing circuits, which are adopted by food spoilage and poisoning bacteria, are still relatively scarce [11]. Undoubtedly, further studies are needed to exploit the effect of other food processing parameters (e.g., temperature, chemical composition) on the expression of the genes that encode signaling molecules. The control of microbial contamination during food processing and storage via conditioning of the quorum sensing systems may fulfill bio-preservative criteria, which are strongly desired by consumers.

5.5   Novel Biogenic Compounds

The molecular language used by probiotic and symbiotic intestinal bacteria was partly decoded. At the intestinal level, bacteria may synthesize, release, detect, and respond to numerous signals, especially low-molecular-mass biogenic compounds of different chemical structure [12]. Global or universal regulators of the interactions between prokaryotic and eukaryotic cells (interkingdom level, see Sect. 4.3) are, therefore, metabolites and structure components of the host and microbial cells, also including their signaling molecules that have, possibly, similar targets. A large number of such biogenic signaling molecules (e.g., lacton-like compounds, peptide pheromones, type AI-2) are reported in the literature, and are distinguished as autoinducers, chemokinins, modulins, or effector molecules [12]. The list also includes compounds such as amino acids (e.g., glutamate and β-alanine), amino acid derivatives (e.g., γ-amino butyric acid and amines), vitamins (e.g., biotin and folic acid) and defensin-like peptides, which are only indirectly related to communication. Overall, the beneficial effects on the host and, especially, on its intestinal microbiota and related immune response seem to be strictly related to the capacity of synthesizing the above compounds at intestinal level. On the basis of these considerations, knowledge concerning the molecular language of probiotic bacteria should allow one to better understand their mode of action and to design novel probiotics with increased health effectiveness. Furthermore, manipulation of the host and its microbiota using tailor-made low-molecular-mass biogenic compounds should be expected to interfere with the cross-talk, stability, and regulation of gene expression.

References

1. Arias CA, Murray BE (2009) Antibiotic-resistant bugs in the 21st century-a clinical super-challenge. N Engl J Med 360:439–443
2. Bjarnsholt T, Jensen PO, Rasmussen TB, Christophersen L, Calum H, Hentzer M, Hougen H-P, Rygaard J, Moser C, Eberl L, Høiby N, Givskov M (2005) Garlic blocks quorum sensing and promotes rapid clearing of pulmonary *Pseudomonas aeruginosa* infections. Microbiology 151:3873–3880
3. Wang LH, Dong YH, Zhang LH (2008) Quorum quenching: impact and mechanisms. In: Winans SC, Bassler BL (eds) Chemical communication among bacteria. ASM Press, Washington, DC, p 379
4. Fast W, Tipton PA (2012) The enzymes of bacterial census and censorship. Trends Biochem Sci 37:7–14

5. Minervini F, De Angelis M, Di Cagno R, Pinto D, Siragusa S, Rizzello CG, Gobbetti M (2010) Robustness of *Lactobacillus plantarum* starters during daily propagation of wheat sourdough type I. Food Microbiol 27:897–908

6. Siragusa S, Di Cagno R, Ercolini D, Minervini F, Gobbetti M, De Angelis M (2009) Taxonomic structure and monitoring of the dominant lactic acid bacteria population during wheat flour sourdough type I propagation by using *Lactobacillus sanfranciscensis* starters. Appl Envir Microbiol 75:1099–1109

7. Di Cagno R, Surico RF, Minervini G, De Angelis M, Rizzello CG, Gobbetti M (2009) Use of autochthonous starters to ferment red and yellow peppers (*Capsicum annum* L.) to be stored at room temperature. Int J Food Microbiol 130:108–116

8. Di Cagno R, Cardinali G, Minervini G, Antonielli L, Rizzello CG, Ricciuti P, Gobbetti M (2010) Taxonomic structure of the yeasts and lactic acid bacteria microbiota of pineapple (*Ananas comosus* L. Merr.) and use of autochthonous starters for minimally processing. Food Microbiol 27:381–389

9. Di Cagno R, Surico RF, Minervini G, Rizzello CG, Lovino R, Servili M, Taticchi A, Urbani S, Gobbetti M (2011) Exploitation of sweet cherry (*Prunus avium* L.) puree added of stem infusion through fermentation by selected autochthonous lactic acid bacteria. Food Microbiol 28:900–909

10. Nychas G-JE, Skandamis PN, Tassou CC, Koutsoumanis K (2008) Meat spoilage during distribution. Meat Sci 78:77–89

11. Blana VA, Doulgeraki A, Nychas G-JE (2011) Autoinducer-2–like activity in lactic acid bacteria isolated from minced beef packaged under modified atmospheres. J Food Prot 74:631–635

12. Shenderov BA (2011) Probiotic (symbiotic) bacterial languages. Anaerobe 17:490–495

Index

M. Gobbetti and R. Di Cagno, *Bacterial Communication in Foods*,
SpringerBriefs in Food, Health, and Nutrition, DOI 10.1007/978-1-4614-5656-8,
© Marco Gobbetti and Raffaella Di Cagno 2013